ABERRATION

AND

THE ELECTROMAGNETIC FIELD.

ABERRATION

AND SOME OTHER PROBLEMS
CONNECTED WITH

THE ELECTROMAGNETIC FIELD.

ONE OF TWO ESSAYS TO WHICH THE ADAMS PRIZE WAS
AWARDED IN 1899, IN THE UNIVERSITY OF CAMBRIDGE.

BY

GILBERT T. WALKER, M.A., B.Sc.
FELLOW AND LECTURER OF TRINITY COLLEGE, CAMBRIDGE.

CAMBRIDGE:
AT THE UNIVERSITY PRESS.
1900

CAMBRIDGE
UNIVERSITY PRESS

University Printing House, Cambridge CB2 8BS, United Kingdom

Cambridge University Press is part of the University of Cambridge.

It furthers the University's mission by disseminating knowledge in the pursuit of education, learning and research at the highest international levels of excellence.

www.cambridge.org
Information on this title: www.cambridge.org/9781107432604

First published 1900
First paperback edition 2014

A catalogue record for this publication is available from the British Library

ISBN 978-1-107-43260-4 Paperback

PREFACE.

THE subject selected by the Examiners for the Adams Prize for 1899 was

THE THEORY OF THE ABERRATION OF LIGHT.

The phenomena of aberration depend upon the relations of the ether and matter and must therefore be intimately associated with many other facts of nature. In the following essay an attempt is made to construct a theory of the electromagnetic field which shall be consistent with the modern interpretation of chemical, optical and magnetic phenomena in terms of electrically charged particles*. That a particular case of such a theory would explain aberration was proved by H. A. Lorentz in the year 1892†.

Since the essay was returned by the examiners I have thought it advisable to remove some defects and obscurities which did not appreciably affect the main argument. It has been chiefly in the latter portion of Part III. that such changes

* A short historical account of electrochemical theories has been given by Richarz (*Phil. Mag.* 39, p. 529, 1895).

† La théorie électromagnétique de Maxwell et son application aux corps mouvants. (*Archives néerlandaises des Sciences exactes et naturelles*, T. xxv.: also published separately by E. J. Brill, Leyden, 1892.)

W. *b*

have been made: and the five pages from (λ), § 56 (p. 78) to the end of Part III. (p. 83) are recent additions for the sake of which the publication has been delayed. I realised last autumn that many of Quincke's classical experiments with magnetic media, on which I had relied for verification, were capable of an interpretation more in accordance with the older theories: and noticing that his investigation of compression* would, under certain conditions, be decisive, I ventured to ask him for some particulars. He has repeated and amplified the experiment† with the conditions modified in order more completely to test the results of § 53. I here wish to express the deepest sense of gratitude to Prof. Quincke for the extremely generous manner in which he has allowed me to avail myself of his great skill.

In Part I. attention has been drawn to an inconsistency in the work of Maxwell, Helmholtz, and others. A medium when magnetised is by them initially regarded as consisting of particles with polar properties—of magnetic 'doublets,' if a hydrodynamical term be permitted. This conception of polarisation will in future be called 'molecular.' On the other hand, when the electromotive force round a moving circuit is determined, the 'induction' or 'magnetic polarisation' is regarded as continuous and completely filling space: it is the change in the number of 'tubes' or 'lines' which is estimated. A similar divergence occurs in relation to electric phenomena. We may follow Maxwell and regard electric displacement as 'continuous'; or we may adopt the point of view of Helmholtz in his paper on anomalous dispersion‡ and suppose that the polarisation of a dielectric consists in a slight disturbance of its ions.

* *Wiedemann's Annalen*, XXIV. p. 380, § 68.

† *Sitzungsber. d. Akad. d. Wiss. z. Berlin*, April 19, 1900.

‡ Elektromagnetische Theorie der Farbenzerstreuung, *Wissenschaft. Abhand.* III. pp. 505—525.

After a study (§§ 3—10) of the geometrical and kinematical properties of a molecularly polarised medium we have formed analytical expressions for Maxwell's fundamental ideas connecting the line-integrals of force round circuits moving in any manner with the fluxes through them. These are the general equations of the field; and those appropriate to molecular polarisation are different from the equations corresponding to continuous polarisation {cf. equations (22) and (27) with (13) and (17)}. The relations between **E**, **H**, the electric and magnetic forces at a point fixed in the ether, and **E**′, **H**′, the forces at a point moving with velocity **u** through it, are also different in the two theories {cf. (14) with (24) and (29)}.

Lorentz in a second paper* adopts the molecular hypothesis, but considers only those cases in which the ether is stationary, the velocity **u** of the matter relative to the ether is constant, and no electric volume-density, conduction currents or magnetic media are present. By a brilliant piece of analysis he transforms (Abschnitt V. §§ 56—59) the equations of the field to those of a stationary system, and obtains (§§ 60—63, 68, &c.) the explanation of the ordinary facts connected with aberration.

The equations of Lorentz are however not sufficiently general for application to such problems as Röntgen's spinning disc, reflection from a rotating mirror, or the determination of stresses in the field.

In Part II. we investigate (§§ 19, 20) a transformation somewhat more general than that of Lorentz already alluded to, and then discuss more closely the propagation of plane waves through a drifting medium. It is found (§§ 21—24) that a medium which is isotropic when stationary, behaves when

* *Versuch einer Theorie der electrischen und optischen Erscheinungen in bewegten Körpern*, Leyden, 1895.

moving like a uniaxal crystal whose axis is parallel to the direction of drift: the difference of velocity of the two waves is of the second order, being

$$\tfrac{1}{2}(K-1)\,\mathbf{u}^2 \sin^2\theta / K^{\frac{3}{2}} V,$$

where θ is the angle between $\mathbf{u}$ and the wave normal.

Röntgen has shown* that a magnetic field is produced by rotating an uncharged glass disc between and parallel to the plates of a charged condenser. The molecular theory yields (§§ 26, 7) a complete explanation.

In Part III. we have investigated the values of the stresses in an electromagnetic field in accordance with several possible hypotheses.

In the determinations of stress made by Maxwell, Helmholtz, Kirchhoff, and Lorberg considerations have been neglected which appear to us of some importance. If we are discussing the magnetic stresses in an electromagnetic field and suppose a displacement effected, there will be changes in the conduction currents and electric forces whose intensities will be proportional to the velocity of the displacement. The result is a rate of redistribution of energy which will be comparable with the rate at which the work is done by the stresses: and it cannot therefore be satisfactory to consider merely the changes in the magnetic potential energy of the system.

Another doubtful method is that of postulating the mechanical effects of the field and then finding stresses which will account for them. This process may be illustrated by considering a magnetostatic field containing permanent magnets but no bodies with susceptibility. If the stresses consist of tensions $\mathbf{H}^2/8\pi$ along lines of force and pressures $\mathbf{H}^2/8\pi$ at right angles to them..(α),

* *Wied. Ann.* XXXV. p. 264, 1888; XL. p. 93, 1890.

the rate at which work is done by the stresses when the velocity at any point is (u, v, w) is, in Maxwell's notation,

$$-\frac{1}{8\pi}\int dv\,\Sigma\,\{u_x(\alpha^2-\beta^2-\gamma^2)+2u_y\alpha\beta+2u_z\alpha\gamma\}.$$

This transforms by Green's theorem to

$$\frac{1}{8\pi}\int dS\,\Sigma u\,\{l(\alpha^2-\beta^2-\gamma^2)+m\,.\,2\alpha\beta+n\,.\,2\alpha\gamma\}_1^2$$

$$+\frac{1}{4\pi}\int dv\,(u\alpha+v\beta+w\gamma)(\alpha_x+\beta_y+\gamma_z),$$

where numbers 1, 2 indicate values inside and outside the magnets. It is easily seen that the surface integral expresses the rate at which work would be done by surface-forces equal to

$$\tfrac{1}{2}(\alpha_1+\alpha_2)\,\upsilon,\quad \tfrac{1}{2}(\beta_1+\beta_2)\,\upsilon,\quad \tfrac{1}{2}(\gamma_1+\gamma_2)\,\upsilon,$$

where υ is the surface density: the volume integral corresponds to an internal force $\alpha\tau$, $\beta\tau$, $\gamma\tau$, per unit volume, where τ is the volume density ....(β).

On replacing τ by $-(A_x+B_y+C_z)$ the rate of doing of work transforms to

$$-\frac{1}{8\pi}\int dS\,\{\alpha^2+\beta^2+\gamma^2\}_1^2\,(lu+mv+nw)$$

$$+\int dv\,\{\Sigma u\,(A\alpha_x+B\alpha_y+C\alpha_z)+\Sigma u_x A\alpha$$

$$+\tfrac{1}{2}\Sigma\,(w_y+v_z)(B\gamma+C\beta)+\tfrac{1}{2}\Sigma\,(w_y-v_z)(B\gamma-C\beta)\}.$$

We now see that the system of stresses (α) is equivalent, in its action on a rigid body, to the system of forces (β), and it is also equivalent to the following system:—

(1) a surface thrust of $-\{\mathbf{H}^2\}_1^2/8\pi$ along the outward normal,

(2) a force in the interior equal to

$$(A\alpha_x+B\alpha_y+C\alpha_z,\quad A\beta_x+B\beta_y+C\beta_z,\quad A\gamma_x+B\gamma_y+C\gamma_z)$$

(i.e. to $\mathbf{I}\nabla\,.\,\mathbf{H}$) per unit volume,

(3) tensions $A\alpha$, $B\beta$, $C\gamma$,

(4) shearing-stresses represented by $\frac{1}{2}(B\gamma + C\beta)$,

(5) couples $B\gamma - C\beta$, &c. ....(γ).

Maxwell states in § 639 of his treatise that the effect of the field is the production of the force (2) and the couples (5), but he does not consider the possibility of the tensions and shears; clearly however the forces on the two poles of a magnetic particle which produce the couples would also produce the tensions (3), and similarly the shears (4) may be interpreted.

Now the system of stresses by means of which Maxwell explains (2) and (5) is given in § 641 as

$$P_{xx} = (\alpha^2 - \beta^2 - \gamma^2)/8\pi + A\alpha, \quad P_{yx} = \alpha\beta/4\pi + B\alpha,$$
$$P_{zx} = \alpha\gamma/4\pi + C\alpha, \text{ \&c.} \qquad (\delta).$$

Thus it is the system (α), together with the tensions (3), shears (4) and couples (5). But the result (β) shows that the system (α) *alone* will explain the forces and couples which act on a rigid magnet taken as a whole. Hence Maxwell's system (δ) is incorrect: for example it will give double as big a resultant couple as it should. This statement may be verified by determining the couple due to the action of a uniform field of magnetic force upon a sphere which is uniformly magnetised and is placed in the field with its direction of magnetisation perpendicular to the lines of force.

The error on which we wish to lay stress is that of using (in § 639) a method which suggests some but not all of the stresses which act on an element of volume: apart from this is the incorrectness of determining stresses in the ether, which fills all space, from an expression for the energy as the volume integral of a function which is zero outside the magnet.

It is found (§§ 34—37) that, if the ether is stationary and

the polarisations of electric and magnetic material media are molecular, the expression

$$\{\mathbf{E}^2 + (K-1)\,\mathbf{E}'^2 + \mathbf{H}^2 + (\mu - 1)\,\mathbf{H}'^2\}/8\pi$$

as the energy per unit volume yields a system of stresses *in the ether* which agrees with Maxwell's in the case of a stationary electrostatic field. It is natural to interpret magnetism as due to the description of small orbits by the ions*, the magnetic moment of a molecule being proportional to the magnetic force acting on it. The relation of the magnetic moment per unit volume to the magnetic force will in that case (§ 47) be different from the relation of the electric moment to the electric force: and the ether-stresses in a magnetic field will be different in character from those in an electric field (§§ 50, 51). On each ion, surrounded as it is by the ether, a resultant force will be produced by the ether-stresses. And these resultant forces acting on the ions, of which the material medium is composed, will call into existence the stresses *within the bodies*; when we know the forces on the component particles of a material medium per unit volume we can find the stress within the body by the application of the ordinary laws of mechanics and of the theory of elasticity (§ 42). Thus we find (§ 52) stresses at surfaces of separation and in the interior of material media which are not Maxwellian, do not involve shears, and are consistent with the equilibrium of a fluid medium.

As Helmholtz pointed out†, Maxwell's theory labours under the disadvantage of giving a finite resultant force

$$\frac{d}{dt}\,[\mathbf{HE}]/4\pi V$$

per unit volume on a stationary element of free ether: in § 50 this difficulty does not arise.

* The extended definition of 'ion' on p. 63 would include the 'corpuscles' recently discussed by J. J. Thomson (*Nature*, May 10, 1900: *Phil. Mag.*, Feb. 1900).

† Helmholtz, *Wiss. Abhand.* III. p. 531.

In §§ 54—56 will be found a comparison of the theoretical stresses of §§ 52, 53 with observations, chiefly on liquid media, made by Quincke and others. The agreement is in all cases as close as could be expected in view of the experimental difficulties.

In Part IV. we have tried to ascertain the possibility of explaining aberration on the hypothesis that the polarisation of material media is continuous, not molecular as is assumed in Part II. §§ 19, 20.

In § 57 it is shown that the velocity of light in a medium drifting with constant velocity u through a stationary ether is increased by an amount $\frac{1}{2}u(K-1)/K$. In § 58 in order to obtain Fresnel's coefficient we are driven to assume that under these conditions the ether is dragged by the material medium with velocity $u(K-1)/(K+1)$: the ordinary facts of aberration are then explained (§§ 59, 60), for the directions of rays prove to be governed by the same laws as in the theory of Lorentz.

As might be expected, Röntgen's experiment with the spinning disc is decisive between the theories of molecular and continuous polarisation. Its explanation on the latter hypothesis is impossible if there is relative motion between the earth and the ether, and this is an inevitable part of the assumption of § 58.

We have availed ourselves of the system of Vector Algebra introduced by Heaviside and adopted by Lorentz and Föppl*: its use appears to give distinctly greater insight as well as greater brevity. A table of vector notation and formulæ is appended, as well as a list of symbols used for expressing the

* Föppl's *Einführung in die Maxwell'sche Theorie der Elektricität* contains in its first section (pp. 5—88) an extremely good introduction to vector algebra. See also Heaviside's *Electromagnetic Theory*, Vol. I. ch. III. 1893.

physical quantities involved. The units employed are those adopted by Hertz*

An attempt has been made to acknowledge indebtedness by giving references whenever assistance has been consciously received: where omissions have inadvertently been made, the largeness of the literature of the subject must be my excuse.

* Hertz, *Ges. Werke*, Vol. II. p. 213: 'Electric Waves,' p. 199.

GILBERT T. WALKER.

TRINITY COLLEGE, CAMBRIDGE,
May 14, 1900.

VECTOR NOTATION AND FORMULAE.

If the components of a vector are denoted by A_1, A_2, A_3 when the vector itself is denoted by $\mathbf{A}$ or by (A_1, A_2, A_3), we have the following scheme :—

$$\mathbf{AB} \equiv \text{the scalar product of } \mathbf{A}, \mathbf{B}$$

$$\equiv A_1B_1 + A_2B_2 + A_3B_3 = \mathbf{BA},$$

$$[\mathbf{AB}] \equiv \text{the vector product of } \mathbf{A}, \mathbf{B}$$

$$\equiv (A_2B_3 - A_3B_2,\ A_3B_1 - A_1B_3,\ A_1B_2 - A_2B_1)$$

$$= -[\mathbf{BA}] \quad \text{..........I,}$$

$$\mathbf{A}\,[\mathbf{BC}] = \mathbf{B}\,[\mathbf{CA}] = \mathbf{C}\,[\mathbf{AB}] = \begin{vmatrix} A_1 & A_2 & A_3 \\ B_1 & B_2 & B_3 \\ C_1 & C_2 & C_3 \end{vmatrix} \quad \text{..........II,}$$

$$[\mathbf{A}\,[\mathbf{BC}]] = \mathbf{B}\,.\,\mathbf{CA} - \mathbf{C}\,.\,\mathbf{AB} \quad \text{..........III,}$$

$$\nabla \equiv \left(\frac{d}{dx}, \frac{d}{dy}, \frac{d}{dz}\right),$$

$$\text{div}\,\mathbf{A} \equiv \nabla\mathbf{A} = \frac{dA_1}{dx} + \frac{dA_2}{dy} + \frac{dA_3}{dz} \quad \text{..........IV,}$$

$$\text{curl}\,\mathbf{A} \equiv [\nabla\mathbf{A}] = \left(\frac{dA_3}{dy} - \frac{dA_2}{dz}, \frac{dA_1}{dz} - \frac{dA_3}{dx}, \frac{dA_2}{dx} - \frac{dA_1}{dy}\right) \quad \text{..........V,}$$

$$\mathbf{A}\nabla\,.\,\mathbf{B} = \left(A_1\frac{d}{dx} + A_2\frac{d}{dy} + A_3\frac{d}{dz}\right)\mathbf{B} \quad \text{..........VI,}$$

$$\text{div}\,[\mathbf{AB}] = \mathbf{B}\,\text{curl}\,\mathbf{A} - \mathbf{A}\,\text{curl}\,\mathbf{B} \quad \text{..........VII,}$$

$$\text{curl}\,[\mathbf{AB}] = \mathbf{A}\,\text{div}\,\mathbf{B} - \mathbf{A}\nabla\,.\,\mathbf{B} - \mathbf{B}\,\text{div}\,\mathbf{A} + \mathbf{B}\nabla\,.\,\mathbf{A} \quad \text{.....VIII,}$$

$$\nabla_B\,.\,\mathbf{AB} \equiv \left(A_1\frac{dB_1}{dx} + A_2\frac{dB_2}{dx} + A_3\frac{dB_3}{dx},\right.$$

$$A_1\frac{dB_1}{dy} + A_2\frac{dB_2}{dy} + A_3\frac{dB_3}{dy},$$

$$\left.A_1\frac{dB_1}{dz} + A_2\frac{dB_2}{dz} + A_3\frac{dB_3}{dz}\right)$$

$$= \mathbf{A}\nabla\,.\,\mathbf{B} + [\mathbf{A}\,\text{curl}\,\mathbf{B}] \quad \text{..........IX.}$$

TABLE OF NOTATION OF PHYSICAL QUANTITIES.

Vector	Components	Meaning
$\mathbf{B}$	a, b, c	Maxwell's 'magnetic induction' $=\mu\mathbf{H}+4\pi\mathbf{I}$.
$\mathbf{C}$	p, q, r	Conduction current.
$\mathbf{D}$	$\mathfrak{X}$, $\mathfrak{Y}$, $\mathfrak{Z}$	$4\pi\times$ (Total electric polarisation exclusive of permanent charges) $=\mathbf{E}+\mathbf{D}'$.
$\mathbf{D}'$	$\mathfrak{X}'$, $\mathfrak{Y}'$, $\mathfrak{Z}'$	$4\pi\times$ (Induced electric polarisation of material medium) $=(K-1)\,\mathbf{E}'$.
$\mathbf{E}$	X, Y, Z	Electric force at a point fixed relative to the ether.
$\mathbf{E}'$	X', Y', Z'	Electric force at a point fixed relative to the moving material medium.
$\mathbf{E}''$	X'', Y'', Z''	Electric force at a point moving with velocity u'', v'', w''.
$\mathbf{F}$	Ξ, H, Z	Mechanical force per unit volume.
$\mathbf{G}$	$\mathfrak{L}$, $\mathfrak{M}$, $\mathfrak{N}$	$4\pi\times$ (Total magnetic polarisation exclusive of permanent magnetisation) $=\mathbf{H}+\mathbf{G}'$.
$\mathbf{G}'$	$\mathfrak{L}'$, $\mathfrak{M}'$, $\mathfrak{N}'$	$4\pi\times$ (Induced magnetic polarisation of the material medium) $=(\mu-1)\,\mathbf{H}'$.
$\mathbf{H}$	L, M, N	Magnetic force at a point fixed relative to the ether.
$\mathbf{H}'$	L', M', N'	Magnetic force at a point fixed relative to the moving material medium.

Vector	Components	Meaning
$\mathbf{H}''$	L'', M'', N''	Magnetic force at a point moving with velocity u'', v'', w''.
$\mathbf{I}$	A, B, C	Intensity of permanent magnetisation.
$\mathbf{M}$	$e\xi, e\eta, e\zeta$	Moment of an electric doublet.
	$m\xi, m\eta, m\zeta$	Moment of a magnetic doublet.
$\mathbf{P}$	P, Q, R	$\frac{1}{4\pi V}[\mathbf{EH}]$.
$\mathbf{P}'$	P', Q', R'	$\frac{1}{4\pi V}[\mathbf{E'H'}]$.
$\mathbf{Q}$	$\mathfrak{P}, \mathfrak{Q}, \mathfrak{R}$	$\frac{1}{4\pi V}[\mathbf{DG}]$.
$\mathbf{u}$	u, v, w	Velocity of matter relative to the free ether.

Components	Meaning
ρ', σ'	Volume and surface densities of the electricity of the induced polarisation.
ρ, σ	Volume and surface densities of the permanent electric charges.
τ', υ'	Volume and surface densities due to induced magnetisation.
$\tau \quad \upsilon$	Volume and surface densities of permanent magnetisation.

CONTENTS.

PART III.

PART IV.

PART I

General Theory.

1. The conception which Faraday introduced of an electromagnetic field as traversed by lines of electric and magnetic force, has, during the last forty years, proved extremely fertile of results.

But the extraordinary success with which Maxwell developed the idea has itself led to the obscuring of several inconsistencies in his work. And it is one of these that delayed the appearance of an explanation, in electromagnetic terms, of the phenomena connected with aberration until the year 1892.

In the statement of the fundamental laws given in his collected papers (Vol. II, p. 138), Maxwell defines the number of lines of magnetic force, whose rate of decrease through a circuit determines the electromotive force round it, as the surface integral of magnetic intensity multiplied by the coefficient of magnetic induction. If the region be mapped out, in the ordinary manner, into tubes of force, the polarisation of a medium of permeability μ may thus be regarded as existing in tubes or lines, and the measure of the polarisation is μ times as great as if the same distribution of tubes of magnetic force existed in empty space.

Again, in his *Recent Researches*, § 8, J. J. Thomson gives a definition from which we quote the following:—

"Let A and B be two neighbouring points in the dielectric, let a plane whose area is unity be drawn between these points and at right angles to the line joining them, then the polarisation in the direction AB is the excess of the number of the tubes which pass through the unit area from the side A to the side B, over those which pass from the side B to the side A."

He shews that if the polarisation be f, g, h and if the tubes be moving with a velocity denoted by u, v, w, then, due to the motion solely, there will be a rate of change given by

$$\frac{df}{dt} = -\left(u\frac{df}{dx} + v\frac{df}{dy} + w\frac{df}{dz}\right) - f\left(\frac{dv}{dy} + \frac{dw}{dz}\right) + g\frac{du}{dy} + h\frac{du}{dz}.$$

On the other hand, in §§ 382–3 of his treatise, Maxwell explains the polarisation of a magnet as consisting in the possession by its particles of equal and opposite properties at the ends of a certain axis through the particle: in fact the particles resemble "doublets" in hydrodynamics.

These two conceptions of polarisation will, in future, be spoken of as "tubular" and "molecular" respectively.

Exactly the same remarks apply to electric polarisation. The "displacement" is equal to the electric force multiplied by $K/4\pi$, and is conceived by Maxwell as forming lines or tubes in space. Increase or diminution of displacement is equivalent to an electric current, and the line integral of magnetic force round any circuit is equal to 4π times the "total" current through it, where by "total" current is meant the increase in the number of tubes of displacement *plus* the conduction current.

On the other hand, the conception of the state of a dielectric as one of "molecular" or "doublet" polarisation is suggested in § 60 of Maxwell's treatise, and has gained enormously in probability through the fact that a modification of it affords an explanation of the anomalous dispersion of light.

It may be easily verified that the equation of electrostatic potential,

$$\frac{d}{dx}\left(K\frac{d\phi}{dx}\right) + \frac{d}{dy}\left(K\frac{d\phi}{dy}\right) + \frac{d}{dz}\left(K\frac{d\phi}{dz}\right) + 4\pi\rho = 0,$$

obtained on the displacement theory follows at once if we suppose that molecular polarisation exists in a dielectric, and that its electric moment per unit volume is $\epsilon\mathbf{E}$ where $K = 1 + 4\pi\epsilon$. This may be deduced in the following manner from the corresponding case in magnetism:—

Consider a medium which has permeability $\mu\,\{=(1 + 4\pi k)\}$

as well as permanent magnetisation whose moment per unit volume is $\mathbf{I}$ $\{\equiv (A, B, C)\}$. According to Maxwell's analysis of a magnetostatic field, the total intensity of magnetisation will then be $k\mathbf{H} + \mathbf{I}$.

The magnetic induction will be $\mathbf{H} + 4\pi(k\mathbf{H} + \mathbf{I})$ or $\mu\mathbf{H} + 4\pi\mathbf{I}$. The solenoidal condition which the induction satisfies is

$$\operatorname{div}(\mu\mathbf{H} + 4\pi\mathbf{I}) = 0,$$

or

$$-\frac{d}{dx}\left(\mu\frac{d\Omega}{dx}\right) - \frac{d}{dy}\left(\mu\frac{d\Omega}{dy}\right) - \frac{d}{dz}\left(\mu\frac{d\Omega}{dz}\right) + 4\pi\left(\frac{dA}{dx} + \frac{dB}{dy} + \frac{dC}{dz}\right) = 0,$$

and hence $$\nabla^2_\mu\Omega + 4\pi\tau = 0 \quad\quad (1),$$

where τ is the volume-density of the permanent magnetism and is equal to $-\operatorname{div}\mathbf{I}$.

Now in a complete electric field the total charge must be zero, and we may therefore replace true charges ρ by a fictitious permanent polarisation $\mathbf{J}$ given by $\operatorname{div}\mathbf{J} + \rho = 0$. If in addition a force $\mathbf{E}$ sets up an "induced" polarisation $\epsilon\mathbf{E}$, an analysis exactly similar to that of the magnetic system will yield the equation

$$\nabla^2_K\phi + 4\pi\rho = 0 \quad\quad (2),$$

where $$K = 1 + 4\pi\epsilon.$$

2. As long as the possibility of relative motion is unconsidered, the difference between the two points of view may remain comparatively unimportant, but with relative motion appear discrepancies, one of which we shall now investigate.

Consider a medium whose polarisation $(\mathfrak{A}, \mathfrak{B}, \mathfrak{C})$ is represented by a number of lines or tubes which are not at rest. It is required to find the rate of change in the number of lines crossing a small circuit, the velocity at any point of which is $\mathbf{u} \equiv (u, v, w)$. Let the area of the circuit be initially α, its shape rectangular, and its plane perpendicular to the axis OX: the number of lines that cross it will at that moment be $\alpha\mathfrak{A}$.

At the end of a short interval of time τ the centre of the circuit will have moved a distance $(u\tau, v\tau, w\tau)$, and the value of $\mathfrak{A}$ will be increased by

$$\tau\left(\frac{d\mathfrak{A}}{dt}+u\frac{d\mathfrak{A}}{dx}+v\frac{d\mathfrak{A}}{dy}+w\frac{d\mathfrak{A}}{dz}\right),$$

or $\tau\frac{d\mathfrak{A}}{dt'}$ say, if $\frac{d}{dt'}\equiv\frac{d}{dt}+\mathbf{u}\nabla$. The circuit is also changing its size and direction. Due to the increase of area

$$\alpha\tau\left(\frac{dv}{dy}+\frac{dw}{dz}\right)$$

there will be an additional number of lines

$$\alpha\tau\left(\frac{dv}{dy}+\frac{dw}{dz}\right)\mathfrak{A}.$$

Due to changes of direction there will be a diminution

$$-\alpha\tau\left(\frac{du}{dy}\mathfrak{B}+\frac{du}{dz}\mathfrak{C}\right).$$

Hence the rate of increase in the number of lines is

$$\alpha\left\{\frac{d\mathfrak{A}}{dt'}+\left(\frac{dv}{dy}+\frac{dw}{dz}\right)\mathfrak{A}-\left(\frac{du}{dy}\mathfrak{B}+\frac{du}{dz}\mathfrak{C}\right)\right\};$$

cf. Hertz, *Ges. Werke*, Bd. II, p. 260; Thomson, *Recent Researches*, pp. 6, 7.

If $(\mathfrak{A}, \mathfrak{B}, \mathfrak{C})\equiv\mathbf{A}$, the rate of change of $\mathbf{A}$ is

$$\frac{d\mathbf{A}}{dt}+\mathbf{u}\nabla\,.\,\mathbf{A}+\mathbf{A}\,.\operatorname{div}\mathbf{u}-\mathbf{A}\nabla\,.\,\mathbf{u},$$

or its equivalent,

$$\frac{d\mathbf{A}}{dt}+\mathbf{u}\operatorname{div}\mathbf{A}+\operatorname{curl}[\mathbf{Au}]\ \ldots\ldots\ldots\ldots\ldots (3).$$

3. Before determining the corresponding flux through a moving area of a substance which is molecularly polarised, some preliminary analysis will be necessary.

It will be assumed throughout that the length of an axis of the molecular "doublet" and the mean distance between the centres of the doublets are each small compared with the small lengths involved in the ordinary determination of the shape of a body by means of differential coefficients with respect to the coordinates. In other words, the lengths of the axes of the doublets and their mean distances apart are small quantities of the second order: in an element of volume, such as would be conceived if we were investigating the density of a body or its conditions of mechanical equilibrium under elastic forces, the number of doublets will be very large.

In the first place, as regards fluxes and potentials, the effect of a doublet is clearly proportional to the strength of its poles. Secondly, if we have a doublet of axis AB, the diagram indicates that by placing pairs of opposite poles at C, D, E equal in magnitude to those at A, B, we shew that a doublet AB may be replaced by the doublets AC, CD, DE, EB; hence, as regards their axes, the doublets obey the vector law.

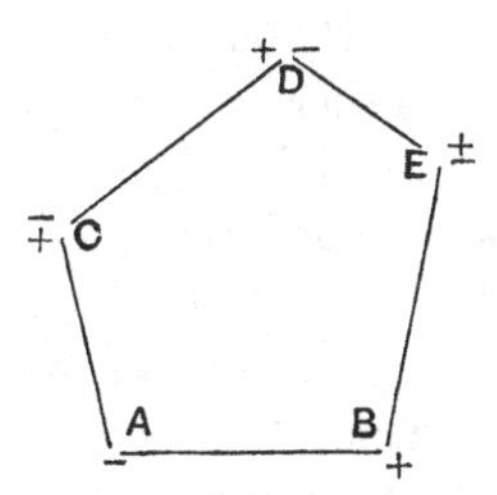

It follows too, from the case in which $ACDEB$ is a straight line, that the effect of a doublet is proportional to the length of its axis as well as to the strength of its poles; that is, the effect varies as the moment. If we replace each doublet by its components parallel to three rectangular axes, we may obtain the effect of a polarised body by adding the effects due to three superposed bodies whose polarisations are respectively parallel to the three rectangular axes. Let us consider one of these, that with its doublets parallel to OX. The intensity of polarisation at any point may be defined by taking a small volume at the point and summing up the moments of the doublets whose centres lie within the volume: the intensity of polarisation is the limit of the total moment divided by the volume.

4. We shall now prove that when the doublets are arranged according to any continuous law, this definition is equivalent to the following:—

If an element of area α be taken, the axes of a certain number n of the doublets will cross the element (i.e. the line joining the poles of any one of these n doublets will cut the plane of the element in a point within the element and between the poles): the intensity of polarisation in the direction of the normal to the element of area will be a quantity $\mathfrak{A}$ such that $\mathfrak{A}\alpha$ is equal to the algebraic sum of the charges on those poles of the n doublets which lie on the same side of the element of area α as the normal in question.

We shall first discuss the case in which all the doublets have axes which are the same in direction and in length. Let this direction be that of OX and let the common length of axis be l. By our former suppositions, a diameter of α will be of the first order of small quantities, while l is of the second order.

Consider the cylinder, of length l and cross section α, which is bounded by two planes parallel to YOZ, distant $\pm\frac{1}{2}l$ from the element of area α, and by generators parallel to OX drawn through points on the margin of the element.

Since the length of the axis of any doublet is equal to the length of this cylinder, it is obvious that the doublets whose centres lie within the cylinder are identical with the doublets whose axes cut the element α. But by the second definition the total strength of the poles on the positive side of these is $\mathfrak{A}\alpha$, and the total moment is $\mathfrak{A}l\alpha$. Hence the sum of the moments of those doublets whose centres lie within the cylinder of volume $l\alpha$ is $\mathfrak{A}l\alpha$.

This cylinder is of length small compared with its diameter: but by combining a number of cylinders of cross section α and length l with their ends in contact we can form a cylinder of length L comparable with its diameter: and to this the first definition is applicable. Its volume will be $L\alpha$ and, the polarisation within it being sensibly uniform, the sum of the moments of the doublets whose centres lie within it will be $\mathfrak{A}L\alpha$. Thus if $\mathfrak{A}$ be defined in the second manner the result is in accordance with the first definition.

When the polarisation consists of doublets whose axes are the same in direction and different in length, the definition holds for those doublets whose length of axis is between l and $l+dl$, where dl is an increment small compared with l. Hence by integration with respect to l we see that the two definitions are equivalent for this case also.

When the directions of the axes of the doublets are different, we replace each of them by its components parallel to the axes. The doublets parallel to OY, OZ will contribute nothing to the sum of the poles on the positive side of an element A parallel to the plane YOZ: and the doublets parallel to OX will contribute $\mathfrak{A}A$ where $\mathfrak{A}$ is the component along OX of the intensity of polarisation according to the former definition.

By taking the element of area and the axis OX in various directions we establish the complete equivalence of the two definitions. We see too that the latter definition gives a quantity obeying the vector law.

Another proof of the equivalence may be deduced from the comparison of equation (15), § 402 of Maxwell's *Electricity and Magnetism*, Vol. II, with his definition of intensity of magnetisation, §§ 384, 385.

5. Consider once more a body whose polarisation is parallel to OX. If the strengths of the poles of a doublet whose centre is at x, y, z are denoted by $\pm m(x, y, z)$, and if the length of the axis is ξ, then summing over a small volume dv, $\Sigma\xi m(x, y, z) = \mathfrak{A}dv$.

Let us first suppose that ξ is constant and m variable through the body. Then at $x'y'z'$ we shall have positive poles of strength $m\left(x' - \frac{\xi}{2}, y', z'\right)$ and negative poles of strength $-m\left(x' + \frac{\xi}{2}, y', z'\right)$; or, in all, $-\xi\frac{d}{dx'}m(x'y'z')$, neglecting squares of ξ.

The total strength within a volume dv will be

$$-\xi\Sigma\frac{d}{dx'}m(x'y'z') \text{ or } -\left\{\frac{d}{dx'}(\mathfrak{A})\right\}dv.$$

The extension of this result to the case in which ξ varies from point to point is a corollary that follows immediately from § 3.

Hence from the three superposed bodies to which a body of polarisation $(\mathfrak{A}, \mathfrak{B}, \mathfrak{C})$ is equivalent, we find that in the general case the volume density is

$$-\frac{d\mathfrak{A}}{dx} - \frac{d\mathfrak{B}}{dy} - \frac{d\mathfrak{C}}{dz}.$$

It is interesting to obtain this result from the second definition, that of § 4.

Consider the region included within a finite closed surface S. The total charge inside S is contributed by those doublets whose axes cross the surface: the equal and opposite poles of a doublet lying altogether within the region will, when taken together, not affect the sum. Now corresponding to an area dS the sum of the poles inside the surface is $(l\mathfrak{A}+m\mathfrak{B}+n\mathfrak{C})dS$ where l, m, n is the inward normal: hence the total charge within the region is $\int dS\,(l\mathfrak{A}+m\mathfrak{B}+n\mathfrak{C})$ or $-\int dv \operatorname{div} \mathbf{A}$, where $\mathbf{A} \equiv (\mathfrak{A}, \mathfrak{B}, \mathfrak{C})$.

Since this result is true for any region whatever, the volume density must be $-\operatorname{div} \mathbf{A}$.

6. In the case of an electrolyte, the application to it of electric force drives the two kinds of ions in opposite directions. We obtain then a portion of fluid in which ions of one sign will preponderate. The mean charge of electricity per unit volume may in such a case, not being zero, be represented by ρ, and if the fluid is in motion with velocity u, v, w, we may describe the motion of the charges as a "convection current" of components ρu, ρv, ρw.

For analytical purposes we may assume that at a point of space there are conduction currents in addition. In that case, the total volume density being θ, if we consider a fixed closed surface S, and denote by p, q, r, the conduction currents, and l, m, n the normal drawn into the volume enclosed by S,

$$\begin{aligned}\int \dot{\theta} dv &= \text{rate of increase of charge}\\ &= \int (lp + mq + nr)\, dS\\ &\quad + \int (l\rho u + m\rho v + n\rho w)\, dS.\end{aligned}$$

But, from Maxwell's Theory, if the displacement is (f, g, h),

$$\int \theta dv = -\int (lf + mg + nh)\, dS.$$

Hence, eliminating θ,

$$\int (lU + mV + nW)\, dS = 0,$$

where $U = \dot{f} + p + \rho u,\ \ V = \dot{g} + q + \rho v,\ \ W = \dot{h} + r + \rho w.$

Therefore over any volume,

$$\int \operatorname{div} \mathbf{U}\, dv = 0,$$

and $\mathbf{U}$ is solenoidal, where $\mathbf{U} \equiv (U, V, W)$.

The similarity between the fundamental relations of electricity and magnetism shews that the total magnetic current, if solenoidal, must include the convection current.

Now, as we shall subsequently see, the equations of the field express in all cases the equality between the total electric or magnetic flow through a circuit, and the curl round that circuit of magnetic or electric force multiplied by a constant depending on the units chosen. But the divergence of the curl of a vector is zero. Hence the divergence of the total electric or magnetic flow must be zero, and the convection currents must be included.

The physical existence of these currents has been verified experimentally by Rowland and others.

7. We shall now investigate the convection currents due to the motion of the magnetic or electric poles which form the doublet polarisation of a medium.

In the first place let us suppose that the axes of the doublets undergo variations, their centres remaining fixed. When the component $\mathfrak{A}$ of the polarisation $(\mathfrak{A}, \mathfrak{B}, \mathfrak{C})$ becomes $\mathfrak{A} + \delta\mathfrak{A}$ this implies that there is a change in the doublets which cross a small area parallel to YOZ: further $\delta\mathfrak{A}$, from the second definition, may be obtained by finding the algebraic sum of those charges which move through the area from its negative to its positive side and subtracting from this sum the algebraic sum of the charges which move across the area in the negative direction. But the result so obtained is clearly the flow of electric or magnetic matter through the area. Hence in this case the rate of flow is $\dfrac{d\mathfrak{A}}{dt}$.

8. Let us consider the flux due to motion with a velocity which is a continuous function of the coordinates.

Suppose a distribution of equal positive magnetic poles of which the volume density is τ_1. If the pole at any point has the velocity $\mathbf{u}$ at that point, then the flux per unit area through any small fixed circuit will be the component along the normal of the vector $\tau_1\mathbf{u}$. If each point of the circuit is moving with a velocity $\mathbf{u}''$, which like $\mathbf{u}$ is a continuous function of the coordinates, the flux through it will be the normal component of $\tau_1(\mathbf{u}-\mathbf{u}'')$.

Suppose that in addition to the above magnetic poles, we have a distribution of poles which are numerically equal to them in magnitude but opposite in sign, and that the volume density of these is $-\tau_2$. Their velocity being $\mathbf{u}$ the flux will now be $(\tau_1-\tau_2)(\mathbf{u}-\mathbf{u}'')$ or $\tau'(\mathbf{u}-\mathbf{u}'')$, where $\tau' (\equiv \tau_1-\tau_2)$ is the resultant volume density.

The distribution of the poles is at present unrestricted: all that we have assumed is that the velocity of the pole at any point is the continuous function $\mathbf{u}$. We are free therefore to suppose that the distribution consists of magnetic doublets.

In that case, however, the fact that the poles of a doublet have the distinct velocities $\mathbf{u}$ of their respective positions will in itself involve a time-rate of change in the moment $\mathbf{M}$ of the doublet. If the poles are of strength $\pm m$, and the axis has projections ξ, η, ζ, then when the centre has moved a small distance $\mathbf{u}\delta t$, the projections of the axis will have become

$$\xi+(\xi u_x+\eta u_y+\zeta u_z)\,\delta t,\ \eta+(\xi v_x+\eta v_y+\zeta v_z)\,\delta t,$$
$$\zeta+(\xi w_x+\eta w_y+\zeta w_z)\,\delta t.$$

Now $\mathbf{M}\equiv(m\xi, m\eta, m\zeta)$ and the moment will thus, in vector notation, have components $m\xi+\delta t\,.\,\mathbf{M}\nabla\,.\,u$, $m\eta+\delta t\,.\,\mathbf{M}\nabla\,.\,v$, $m\zeta+\delta t\,.\,\mathbf{M}\nabla\,.\,w$: hence $\mathbf{M}$ will be changing at a rate $\mathbf{M}\nabla\,.\,\mathbf{u}$.

Let the intensity of magnetisation be $\mathbf{G}'/4\pi$: then 4π times the moment of the doublets within a small volume ω will be $\omega\mathbf{G}'$: and if the poles of all the doublets have their several velocities $\mathbf{u}$, the doublets that were in the volume ω will after a short time δt have a moment

$$\omega\,(\mathbf{G}'+\delta t\,.\,\mathbf{G}'\nabla\,.\,\mathbf{u})/4\pi.$$

Also $4\pi\tau' = -\operatorname{div} \mathbf{G}'$, and 4π times the consequent flux through a circuit moving with velocity $\mathbf{u}''$ will be the component normal to the circuit of $(\mathbf{u}'' - \mathbf{u}) \operatorname{div} \mathbf{G}'$.

9. Hence if, by means of § 7, we correct this flux for the changes in the moments of the individual doublets we arrive at the following conclusion :—

If the centres of the doublets are displaced with velocity $\mathbf{u}$ and their individual moments are not altered in magnitude or direction (i.e. each doublet is moved as a rigid body without rotation), then 4π times the flux through a circuit moving with velocity $\mathbf{u}''$ will be the normal component of

$$(\mathbf{u}'' - \mathbf{u}) \operatorname{div} \mathbf{G}' - \mathbf{G}'\nabla . \mathbf{u} \ldots\ldots\ldots\ldots\ldots(5),$$

Now $\mathbf{G}'$ represents, by definition, 4π times the moment per unit volume: and the doublets which occupy volume ω will at the end of an interval δt occupy volume $\omega (1 + \delta t . \operatorname{div} \mathbf{u})$, their total moment, according to § 9, remaining unaltered. Thus the mere change in the relative position of the doublets will add to $\mathbf{G}'$ an amount $\delta\mathbf{G}'$ given by

$$\omega (1 + \delta t \operatorname{div} \mathbf{u}) (\mathbf{G}' + \delta\mathbf{G}') = \omega\mathbf{G}' ;$$

and $$\delta\mathbf{G}' = - \delta t . \mathbf{G}' \operatorname{div} \mathbf{u}.$$

If then the moment per unit volume is to remain unaltered we must allow for the geometrical expansion by giving $\mathbf{G}'$ a time-rate $\mathbf{G}' \operatorname{div} \mathbf{u}$. This will yield a corresponding flux, and from (5) we deduce the conclusion following :—

If the medium moves with velocity $\mathbf{u}$ and the moment per unit volume remains unaltered, then 4π times the flux through a circuit moving with velocity $\mathbf{u}''$ will be the normal component of

$$(\mathbf{u}'' - \mathbf{u}) \operatorname{div} \mathbf{G}' - \mathbf{G}'\nabla . \mathbf{u} + \mathbf{G}' \operatorname{div} \mathbf{u} \ \ldots\ldots\ldots(6).$$

10. Another consequence of the change of volume may be conveniently noted at this point.

We can conceive two modes in which the susceptibility of a medium may be conceived. According to the ordinary definition, the susceptibility is the ratio of the magnetic moment of the medium *per unit volume* to the magnetic force.........(α).

On the other hand we might have a medium so constituted that it is the individual magnetic doublets whose moments bear constant ratios to the magnetic forces acting on them ...(β).

In the case of media at rest, the distinction between the two conceptions is unimportant. But if a medium of the second type undergoes distortion, the susceptibility, as defined in the ordinary way and dependent on the moment per unit volume, will undergo variation.

Let a medium of this second type be subject to magnetic force $\mathbf{H}'$, and let $\mathbf{G}'/4\pi$, its moment per unit volume, be equal to $k\mathbf{H}'$, so that k is the susceptibility according to the usual definition. Suppose, as in § 9, that the centres of the doublets are moving with velocity $\mathbf{u}$ and that their individual moments bear a ratio to $\mathbf{H}'$ which remains constant. Then the doublets which at the commencement of the short interval of time δt occupied a volume ω had then a moment $k\mathbf{H}'\omega$. After the interval δt the moment of these doublets will be $k\mathbf{H}_1'\omega$, where $\mathbf{H}_1'$ represents the value of $\mathbf{H}'$ after the interval. But the volume occupied by them will be $\omega(1 + \delta t \,.\, \text{div}\,\mathbf{u})$.

Hence the moment per unit volume will bear to $\mathbf{H}_1'$ the ratio $k + \delta k$, where

$$(k + \delta k)\,\mathbf{H}_1'\omega\,(1 + \delta t \,.\, \text{div}\,\mathbf{u}) = k\mathbf{H}_1'\omega,$$

i.e.

$$\delta k = -\,k\delta t \,.\, \text{div}\,\mathbf{u}.$$

Thus the distortion causes in the susceptibility, if defined in the ordinary manner, a time-rate of change given by

$$\frac{dk}{dt'} = -\,k\,\text{div}\,\mathbf{u}.$$

Hence too

$$\frac{d\mu}{dt'} = -\,(\mu - 1)\,\text{div}\,\mathbf{u} \quad \text{..................(7).}$$

11. We shall now consider the electric forces in action upon a conducting wire which moves through a magnetic field.

If the magnetic induction $\mathbf{B}$ is regarded as occurring in lines or tubes, and if there is no distribution of permanently magnetised matter in the neighbourhood of the wire, then

Maxwell's treatise gives as the expression of Faraday's Law (§ 598)

$$P = cv - bw - \frac{dF}{dt} - \frac{d\psi}{dx},$$

or

$$\mathbf{E}' = \mathbf{E} + [\mathbf{uB}],$$

where $\mathbf{E}'$ is the force on the moving conductor.

In his investigation of the forces in a moving body, Hertz follows Faraday in equating the line integral of electric force round a circuit, moving at each point with the matter, to $4\pi/V$ times the rate of decrease of the number of lines of magnetic polarisation that cross the circuit. In his expression for the rate of decrease he does not explicitly introduce any rate of change due to motion on the part of lines of induction of the kind pointed out by J. J. Thomson (*Recent Researches*, § 9). The rate of variation due to such motion would be included by Hertz in the differential coefficient of the polarisation with respect to the time.

Hertz finds from the consideration of a moving circuit, in accordance with Maxwell's ideas (Maxwell's Papers, II, p. 138), that if $\mathbf{E}'$, $\mathbf{H}'$ be the forces at points in the moving body, and $(\mathfrak{X}, \mathfrak{Y}, \mathfrak{Z})$, $(\mathfrak{L}, \mathfrak{M}, \mathfrak{N})$ be the complete electric and magnetic polarisations,

$$\frac{d\mathfrak{L}}{dt} + \mathfrak{L}\left(\frac{dv}{dy} + \frac{dw}{dz}\right) - \mathfrak{M}\frac{du}{dy} - \mathfrak{N}\frac{du}{dz} = V\left(\frac{dY'}{dz} - \frac{dZ'}{dy}\right).$$

In vector form, if $(\mathfrak{X}, \mathfrak{Y}, \mathfrak{Z}) \equiv \mathbf{D}$ and $(\mathfrak{L}, \mathfrak{M}, \mathfrak{N}) \equiv \mathbf{G}$,

$$\frac{d\mathbf{G}}{dt} + \mathbf{u}\nabla \,.\, \mathbf{G} + \mathbf{G}\,\mathrm{div}\,\mathbf{u} - \mathbf{G}\nabla \,.\, \mathbf{u} = -V\,\mathrm{curl}\,\mathbf{E}' \quad \text{...(8)}.$$

This equation will sometimes be written in the form

$$\frac{\delta\mathbf{G}}{\delta t} = -V\,\mathrm{curl}\,\mathbf{E}' \quad \text{..................(9)}.$$

But

$$\mathbf{G}\,\mathrm{div}\,\mathbf{u} - \mathbf{G}\nabla \,.\, \mathbf{u} - \mathbf{u}\,\mathrm{div}\,\mathbf{G} + \mathbf{u}\nabla \,.\, \mathbf{G} \equiv \mathrm{curl}\,[\mathbf{Gu}],$$

$$\therefore\ \frac{d\mathbf{G}}{dt} + \mathbf{u}\,\mathrm{div}\,\mathbf{G} + \mathrm{curl}\,[\mathbf{Gu}] = -V\,\mathrm{curl}\,\mathbf{E}',$$

or

$$\frac{d\mathbf{G}}{dt} + \mathbf{u}\,\mathrm{div}\,\mathbf{G} = -V\,\mathrm{curl}\left(\mathbf{E}' - \frac{1}{V}[\mathbf{uG}]\right) \quad \text{......(10)}.$$

Similarly, if the conduction current is $\mathbf{C}$,

$$\frac{\delta \mathbf{D}}{\delta t} + 4\pi \mathbf{C} = V \operatorname{curl} \mathbf{H}' \qquad \text{.................(11)},$$

or
$$\frac{d\mathbf{D}}{dt} + \mathbf{u} \operatorname{div} \mathbf{D} + 4\pi \mathbf{C} = V \operatorname{curl} \left(\mathbf{H}' + \frac{1}{V}[\mathbf{uD}]\right) \ldots(12).$$

Now $\mathbf{H}$, $\mathbf{E}$ are forces exerted on a point at rest, and if we consider a circuit fixed in space, we must include the convection currents $\mathbf{u}\rho$, $\mathbf{u}\tau$ (where ρ, τ are the permanent electric and magnetic volume densities as distinct from ρ', τ' due to the induced polarisations). Thus

$$\left.\begin{aligned} \frac{d\mathbf{G}}{dt} + 4\pi \mathbf{u}\tau &= -V \operatorname{curl} \mathbf{E} \\ \frac{d\mathbf{D}}{dt} + 4\pi \mathbf{u}\rho + 4\pi \mathbf{C} &= V \operatorname{curl} \mathbf{H} \end{aligned}\right\} \qquad \text{............(13)}.$$

The theories of Maxwell and Hertz (*Ges. Werke*, II, pp. 226, 265 : 1895) give us $\operatorname{div} \mathbf{D} = 4\pi\rho$; and similarly, $\operatorname{div} \mathbf{G} = 4\pi\tau$. Hence, comparing (10), (12), (13),

$$\left.\begin{aligned} \mathbf{E}' &= \mathbf{E} + \frac{1}{V}[\mathbf{uG}] \\ \mathbf{H}' &= \mathbf{H} - \frac{1}{V}[\mathbf{uD}] \end{aligned}\right\} \qquad \text{...................(14)}.$$

These analytical results are capable of a geometrical and also of a physical interpretation.

In the first place if $\mathbf{E}$, $\mathbf{H}$, $\mathbf{D}$, $\mathbf{G}$ are regarded merely as vectors connected by the relations implied in (13), then we have geometrical connections between the line integrals round a fixed circuit of the tangential components of $\mathbf{E}$, $\mathbf{H}$ and the fluxes of certain quantities through the circuit. The equations (9), (11) shew that corresponding geometrical relations exist when the circuit is moving with velocity $\mathbf{u}$ at each point, provided that we take the line-integral of $\mathbf{E}'$, $\mathbf{H}'$ (as defined by (14)) instead of $\mathbf{E}$, $\mathbf{H}$.

The fact that $\mathbf{E}'$, $\mathbf{H}'$ are vectors, and are at each point independent of the direction of the circuit at that point, suggests the physical interpretation (cf. Maxwell, § 598). Since the electric or magnetic force along the tangent to the *moving* circuit is the component of $\mathbf{E}'$ or $\mathbf{H}'$ obtained by resolving along that tangent, it is natural to regard $\mathbf{E}'$, $\mathbf{H}'$ as the forces which would act on a particle moving with velocity $\mathbf{u}$ and possessing a unit charge of electricity or of magnetism.

In order to distinguish $\mathbf{E}'$, $\mathbf{H}'$ from $\mathbf{E}$, $\mathbf{H}$, which give the force acting on a particle fixed in the ether, we shall call $\mathbf{E}'$, $\mathbf{H}'$ the "forces at a point moving with velocity $\mathbf{u}$."

12. Let us now investigate the forces $\mathbf{E}''$, $\mathbf{H}''$ at a point moving with velocity u'', v'', w''. Considering a small circuit whose velocity at each point is the value of $\mathbf{u}''$ at that point, we find as the increase in the number of lines of electric polarisation,

$$\frac{d\mathbf{D}}{dt} + \mathbf{u}''\nabla \, . \, \mathbf{D} + \mathbf{D} \operatorname{div} \mathbf{u}'' - \mathbf{D}\nabla \, . \, \mathbf{u}'',$$

and hence, adding the conduction current and a convection current through the circuit of amount $(\mathbf{u} - \mathbf{u}'')\,\rho$, we have

$$\frac{d\mathbf{D}}{dt} + \mathbf{u}'' \operatorname{div} \mathbf{D} + \operatorname{curl} [\mathbf{D}\mathbf{u}''] + 4\pi \left\{(\mathbf{u} - \mathbf{u}'')\,\rho + \mathbf{C}\right\} = V \operatorname{curl} \mathbf{H}'',$$

$$\therefore \; \frac{d\mathbf{D}}{dt} + \mathbf{u} \operatorname{div} \mathbf{D} + \operatorname{curl} [\mathbf{D}\mathbf{u}''] + 4\pi\mathbf{C} = V \operatorname{curl} \mathbf{H}''.$$

But $$\frac{d\mathbf{D}}{dt} + \mathbf{u} \operatorname{div} \mathbf{D} + 4\pi\mathbf{C} = V \operatorname{curl} \mathbf{H}.$$

Hence all the conditions are satisfied by the value

$$\mathbf{H}'' = \mathbf{H} - \frac{1}{V}[\mathbf{u}''\mathbf{D}] \;\;\ldots\ldots\ldots\ldots\ldots\ldots(15).$$

Similarly, in order that the line-integral of electric force round a circuit moving in any manner may be equal to -4π times the *total* magnetic current through it,

$$\frac{d\mathbf{G}}{dt} + \mathbf{u} \operatorname{div} \mathbf{G} + \operatorname{curl} [\mathbf{G}\mathbf{u}''] = - V \operatorname{curl} \mathbf{E}'',$$

$$\therefore \; \mathbf{E}'' = \mathbf{E} + \frac{1}{V}[\mathbf{u}''\mathbf{G}] \;\;\ldots\ldots\ldots\ldots\ldots\ldots(16).$$

As in the previous section, we interpret $\mathbf{E}''$, $\mathbf{H}''$ as the forces at a point moving with velocity $\mathbf{u}''$.

13. In the case of a piece of iron magnetised by induction, if the ether and the matter are at rest, it is natural to speak of the polarisation as consisting of $\mathbf{H}/4\pi$ in the ether and $(\mu - 1)\,\mathbf{H}/4\pi$ in the matter. So in general, whether the ether and the matter are at rest or moving, the electric and the magnetic polarisation may be each regarded as consisting of two portions, one in the ether and the other in the matter. After obtaining in § 598 the electric force (or, as Maxwell calls it, the "electromotive force") on a moving element, he says:—

"If the body is a conductor, the electromotive force will produce a current; if it is a dielectric, the electromotive force will produce only electric displacement."

If $\mathbf{E}'$ is the electric force that would tend to set up conduction currents in the body, it would be gratuitously paradoxical to assume that 4π times the polarisation in the body, instead of being $(K-1)\,\mathbf{E}'$, were $(K-1)\mathbf{E}$ or $(K-1)\,\mathbf{E}''$.

Similarly the polarisation in the ether will be equal to the force on an element fixed in the ether. Thus if the ether is at rest, and the matter has velocity u,

$$\left.\begin{aligned} \mathbf{G} &= \mathbf{H} + (\mu - 1)\,\mathbf{H}' \\ \mathbf{D} &= \mathbf{E} + (K - 1)\,\mathbf{E}' \end{aligned}\right\} \quad \text{..................} \quad (17).$$

If the ether and the body have the same velocity u,

$$\left.\begin{aligned} \mathbf{G} &= \mathbf{H}' + (\mu - 1)\,\mathbf{H}' = \mu\mathbf{H}' \\ \mathbf{D} &= \mathbf{E}' + (K - 1)\,\mathbf{E}' = K\mathbf{E}' \end{aligned}\right\} \quad \text{............} \quad (18),$$

as with Hertz.

14. We shall follow Maxwell in regarding the polarisation in the ether as "continuous" or "tubular": but the polarisation in the material medium will now be regarded as occurring in doublets.

Let the ether be treated as stationary and the material medium as having velocity $\mathbf{u}$: then 4π times the flux of the electric polarisation in the ether through a circuit moving with velocity $\mathbf{u}''$ will be

$$\frac{d\mathbf{E}}{dt''} + \mathbf{E}\,\mathrm{div}\,\mathbf{u}'' - \mathbf{E}\nabla\,.\,\mathbf{u}'':$$

here

$$\frac{d}{dt''} \equiv \frac{d}{dt} + u''\frac{d}{dx} + v''\frac{d}{dy} + w''\frac{d}{dz} \equiv \frac{d}{dt} + \mathbf{u}''\nabla.$$

We have seen too, that if the intensity $\mathbf{D}'$ of polarisation of the medium remain unaltered, 4π times its contribution to the flux will be

$$(\mathbf{u}'' - \mathbf{u}) \operatorname{div} \mathbf{D}' + \mathbf{D}' \operatorname{div} \mathbf{u} - \mathbf{D}'\nabla . \mathbf{u}.$$

But $4\pi\mathbf{D}' = (K-1)\mathbf{E}'$, so that instead of being constant $\mathbf{D}'$ will have a time rate $\frac{K-1}{4\pi}\frac{d\mathbf{E}'}{dt'}$: and this quantity must be added to obtain the actual flux. Hence the flux of polarisation multiplied by 4π is

$$\frac{d\mathbf{E}}{dt''} + \mathbf{E} \operatorname{div} \mathbf{u}'' - \mathbf{E}\nabla . \mathbf{u}'' + (K-1)\left\{\frac{d\mathbf{E}'}{dt'} + \mathbf{E}' \operatorname{div} \mathbf{u} - \mathbf{E}'\nabla . \mathbf{u} + (\mathbf{u}'' - \mathbf{u}) \operatorname{div} \mathbf{E}'\right\}.$$

For the total flux we must add the conduction current $\mathbf{C}$ and the convection current due to any volume density ρ of electricity independent of the polarisation; this will be regarded as lodged in the medium and moving with it. The consequent addition will be

$$\mathbf{C} + (\mathbf{u} - \mathbf{u}'')\rho$$

where

$$\operatorname{div} \mathbf{E} = 4\pi\,(\text{total density}) = 4\pi(\rho' + \rho)$$
$$= -\operatorname{div} \mathbf{D}' + 4\pi\rho;$$

$$\therefore \operatorname{div} \mathbf{D} = 4\pi\rho \qquad \text{.. (19).}$$

Thus we shall have

$$\frac{d\mathbf{E}}{dt''} + \mathbf{E} \operatorname{div} \mathbf{u}'' - \mathbf{E}\nabla . \mathbf{u}'' + (K-1)\left\{\frac{d\mathbf{E}'}{dt'} + \mathbf{E}' \operatorname{div} \mathbf{u} - \mathbf{E}'\nabla . \mathbf{u} + (\mathbf{u}'' - \mathbf{u}) \operatorname{div} \mathbf{E}'\right\} + 4\pi\mathbf{C} + 4\pi(\mathbf{u} - \mathbf{u}'')\rho = V \operatorname{curl} \mathbf{H}'',$$

or,

$$\frac{d\mathbf{E}}{dt''} + \mathbf{E} \operatorname{div} \mathbf{u}'' - \mathbf{E}\nabla . \mathbf{u}'' + (\mathbf{u} - \mathbf{u}'') \operatorname{div} \mathbf{E} + (K-1)\left\{\frac{d\mathbf{E}'}{dt'} + \mathbf{E}' \operatorname{div} \mathbf{u} - \mathbf{E}'\nabla . \mathbf{u}\right\} + 4\pi\mathbf{C} = V \operatorname{curl} \mathbf{H}'' \qquad \text{............(20).}$$

Taking $\mathbf{u}''$ as $\mathbf{u}$ we have, on simplifying,

$$\frac{d\mathbf{D}}{dt'} + \mathbf{D}\operatorname{div}\mathbf{u} - \mathbf{D}\nabla\,.\,\mathbf{u} + 4\pi\mathbf{C} = V\operatorname{curl}\mathbf{H}' \quad\text{......} (21)$$

as in the case in which the material polarisation was "continuous," not "molecular."

Taking $\mathbf{u}''$ as zero, we have

$$\frac{d\mathbf{E}}{dt} + \mathbf{u}\operatorname{div}\mathbf{E} + (K-1)\left\{\frac{d\mathbf{E}'}{dt'} + \mathbf{E}'\operatorname{div}\mathbf{u} - \mathbf{E}'\nabla.\mathbf{u}\right\}$$
$$+ 4\pi\mathbf{C} = V\operatorname{curl}\mathbf{H} \quad\text{.........}(22).$$

Thus

$$V\operatorname{curl}(\mathbf{H}'' - \mathbf{H}) = \mathbf{u}''\nabla\,.\,\mathbf{E} + \mathbf{E}\operatorname{div}\mathbf{u}'' - \mathbf{E}\nabla\,.\,\mathbf{u}'' - \mathbf{u}''\operatorname{div}\mathbf{E}$$
$$= \operatorname{curl}[\mathbf{E}\mathbf{u}''], \text{ (by formula VIII)}$$

leading to

$$\mathbf{H}'' = \mathbf{H} + \frac{1}{V}[\mathbf{E}\mathbf{u}''] \quad\text{.................} (23).$$

Similarly

$$\mathbf{H}' = \mathbf{H} + \frac{1}{V}[\mathbf{E}\mathbf{u}] \quad\text{.................} (24).$$

If we regard the magnetic polarisation of the material medium as related to the magnetic force in exactly the same manner as is the case with electric polarisation, we obtain the corresponding equations :—

$$\frac{d\mathbf{H}}{dt''} + \mathbf{H}\operatorname{div}\mathbf{u}'' - \mathbf{H}\nabla\,.\,\mathbf{u}'' + (\mathbf{u} - \mathbf{u}'')\operatorname{div}\mathbf{H}$$
$$+ (\mu - 1)\left\{\frac{d\mathbf{H}'}{dt'} + \mathbf{H}'\operatorname{div}\mathbf{u} - \mathbf{H}'\nabla.\mathbf{u}\right\} = -V\operatorname{curl}\mathbf{E}'' \quad\text{...}(25)$$

$$\frac{d\mathbf{G}}{dt'} + \mathbf{G}\operatorname{div}\mathbf{u} - \mathbf{G}\nabla\,.\,\mathbf{u} = -V\operatorname{curl}\mathbf{E}' \quad\text{...} (26)$$

$$\frac{d\mathbf{H}}{dt} + \mathbf{u}\operatorname{div}\mathbf{H} + (\mu - 1)\left\{\frac{d\mathbf{H}'}{dt'} + \mathbf{H}'\operatorname{div}\mathbf{u} - \mathbf{H}'\nabla.\mathbf{u}\right\}$$
$$= -V\operatorname{curl}\mathbf{E} \quad\text{...} (27)$$

$$\operatorname{div}\mathbf{G} = 4\pi\tau \quad\text{...........}(28).$$

Hence

$$\mathbf{E}'' = \mathbf{E} + \frac{1}{V}[\mathbf{u}''\mathbf{H}], \text{ and } \mathbf{E}' = \mathbf{E} + \frac{1}{V}[\mathbf{u}\mathbf{H}] \quad\text{...}(29).$$

15. If the polarisation of the stationary ether as well as that of the moving material medium had been supposed to be "molecular" we should have obtained the equations:

$$\frac{d\mathbf{E}}{dt}+\mathbf{u}''\operatorname{div}\mathbf{E}+(K-1)\left\{\frac{d\mathbf{E}'}{dt'}+(\mathbf{u}''-\mathbf{u})\operatorname{div}\mathbf{E}'+\mathbf{E}'\operatorname{div}\mathbf{u}-\mathbf{E}'\nabla.\mathbf{u}\right\}$$
$$+4\pi(\mathbf{u}-\mathbf{u}'')\rho+4\pi\mathbf{C}=\quad V\operatorname{curl}\mathbf{H}'',$$

reducing to

$$\frac{d\mathbf{E}}{dt}+\mathbf{u}\operatorname{div}\mathbf{E}+(K-1)\left\{\frac{d\mathbf{E}'}{dt'}+\mathbf{E}'\operatorname{div}\mathbf{u}-\mathbf{E}'\nabla.\mathbf{u}\right\}$$
$$+4\pi\mathbf{C}=\quad V\operatorname{curl}\mathbf{H}'',$$

and

$$\frac{d\mathbf{H}}{dt}+\mathbf{u}\operatorname{div}\mathbf{H}+(\mu-1)\left\{\frac{d\mathbf{H}'}{dt'}+\mathbf{H}'\operatorname{div}\mathbf{u}-\mathbf{H}'\nabla\mathbf{u}\right\}=-V\operatorname{curl}\mathbf{E}''.$$

On replacing $\mathbf{u}''$ by $\mathbf{u}$ or by zero, the left sides of the last two equations remain unaffected, and it is seen that the above hypothesis leads to

$$\left.\begin{aligned}\mathbf{E}''=\mathbf{E}'=\mathbf{E}\\ \mathbf{H}''=\mathbf{H}'=\mathbf{H}\end{aligned}\right\}\quad\text{......................(30).}$$

16. We have now three hypotheses leading to the three equations (§ 11, (14); § 14, (29); § 15, (30))

$$\mathbf{E}'=\mathbf{E}+\frac{1}{V}[\mathbf{uG}]\quad\text{......................}(\alpha),$$

$$\mathbf{E}'=\mathbf{E}+\frac{1}{V}[\mathbf{uH}]\quad\text{...................}(\beta),$$

$$\mathbf{E}'=\mathbf{E}\quad\text{.................................}(\gamma),$$

of which the last is certainly contradicted by the experiment of observing the electric force on a wire moving through a magnetic field.

In a discussion of the mechanical force on a piece of iron carrying an electric current (*Phil. Mag.* 46, p. 154, 1898), J. J. Thomson shews that both (α) and (β) may be reconciled with Maxwell's theory.

As we shall see later, the hypothesis which leads to (α) is negatived by the facts connected with aberration and with several other phenomena.

It will be noticed that hitherto no assumption has been necessary as to the rest or motion of the tubes of force. We have indicated by $\frac{d}{dt}$ and $\frac{d}{dt'}$ the time rates of change at a point at rest and at a point moving with velocity $\mathbf{u}$, and any changes in the polarisation due to possible motion have been supposed included in the operations $\frac{d}{dt}, \frac{d}{dt'}$. Hence it follows that though Maxwell's discussion of the force on a moving conductor is based on the supposition that the tubes of induction are at rest, and Hertz believed that the ether had at each point the same velocity as the matter there present, their equations are reconcileable in form.

Boundary conditions.

17. Let us consider a surface of separation between two media which have a common velocity $\mathbf{u}$ at the interface. The polarisation in the ether will be regarded as "tubular": then the equations (9), (12) of § 11 for "tubular" polarisation of the material medium are identical in form with (21) and (26) of § 14 for "molecular" polarisation of the material medium. And as it is on these equations that the following analysis depends, the results obtained will be independent of the nature of the material media in question.

Let us take the origin in the bounding surface and OZ as drawn along the normal into the second medium.

The surface of discontinuity may be regarded as the limiting case of an indefinitely thin transition-layer in which the quantities involved change rapidly but continuously (cf. Hertz, *Ges. Werke*, II, pp. 220—223, § 8, pp. 271, 2, § 5).

Within the region of change the forces $\mathbf{E}, \mathbf{H}, \mathbf{E}', \mathbf{H}'$ may be regarded as finite, but differential coefficients with regard to z may become very large when the transition-layer is very thin. Since $\mathbf{u}$ is continuous at the boundary its differential coefficients with respect to x, y, z will remain finite.

The equations within the region of transition will be, if $\mathbf{D} \equiv (\mathfrak{X}, \mathfrak{Y}, \mathfrak{Z})$ as before,

$$\frac{d\mathfrak{X}}{dt'} + \mathfrak{X} \operatorname{div} \mathbf{u} - \mathbf{D}\nabla . u + 4\pi p = V(N'_y - M'_z) \ldots (31),$$

$$\frac{d\mathfrak{Y}}{dt'} + \mathfrak{Y} \operatorname{div} \mathbf{u} - \mathbf{D}\nabla . v + 4\pi q = V(L'_z - N'_x) \ldots (32),$$

$$\frac{d\mathfrak{Z}}{dt'} + \mathfrak{Z} \operatorname{div} \mathbf{u} - \mathbf{D}\nabla . w + 4\pi r = V(M'_x - L'_y) \ldots (33),$$

with

$$\operatorname{div} \mathbf{D} = 4\pi\rho \ldots\ldots\ldots\ldots (34),$$

and a corresponding set of magnetic equations.

If we integrate the first equation with respect to z along a line which passes from the first medium into the second, we have

$$\int_1^2 dz \left\{ \frac{d\mathfrak{X}}{dt'} + \mathfrak{X} \operatorname{div} \mathbf{u} - \mathbf{D}\nabla . u + 4\pi p - VN'_y \right\} = - V \{M'\}_1^2,$$

and the expression to be integrated is finite. Hence in the limit, when the layer of transition is extremely thin,

$$\{M'\}_1^2 = 0 \ldots\ldots\ldots\ldots\ldots\ldots\ldots\ldots (35).$$

Similarly from the second equation

$$\{L'\}_1^2 = 0 \ldots\ldots\ldots\ldots\ldots\ldots\ldots\ldots (36).$$

Then $(M'_x - L'_y)$ having equal values on opposite sides of the interface, the third equation shews that

$$\left\{ \frac{d\mathfrak{Z}}{dt'} + \mathfrak{Z} \operatorname{div} \mathbf{u} - \mathbf{D}\nabla . w + 4\pi r \right\}_1^2 = 0 \ldots\ldots\ldots (37),$$

and the fourth, on integrating to z,

$$\{\mathfrak{Z}\}_1^2 = 4\pi\sigma \ldots\ldots\ldots\ldots\ldots\ldots\ldots\ldots (38),$$

where $\sigma \equiv$ the surface density of electricity.

It will be noticed however that these conditions are not independent. If the tangential component of $\mathbf{H}'$ is continuous and the equations which hold within the two media are satisfied, then (37) is a corollary. It is possible also to regard (38) rather as a definition of surface density than as a new physical law.

The magnetic equations give in a similar manner, if

$\mathbf{G} \equiv (\mathfrak{L}, \mathfrak{M}, \mathfrak{N})$ as before,

$$\{X'\}_1^2 = 0, \quad \{Y'\}_1^2 = 0 \quad \text{.................}(39),$$

$$\left\{\frac{d\mathfrak{N}}{dt'} + \mathfrak{N}\,\operatorname{div}\mathbf{u} - \mathbf{G}\nabla \,.\, w\right\}_1^2 = 0 \quad \text{...........}(40),$$

$$\{\mathfrak{N}\}_1^2 = 4\pi\upsilon \quad \text{........................}(41),$$

where $\upsilon \equiv$ the surface density due to permanent magnetism. As before, (40) is not an independent condition.

18. These conditions may be applied to the case in which $K=1$, $\mu=1$ in the second medium: we shall then be dealing with free ether in that region.

If the first medium is a conductor, and the conditions are "steady" with no conduction currents existing, then within the conductor the force $\mathbf{E}'$, which tends to set up currents, must vanish. Hence just outside the conductor, the tangential component of $\mathbf{E}'$ must vanish, and the direction of $\mathbf{E}'$ must be normal to the surface. (Cf. Heaviside, *Electrical Papers*, Vol. II, p. 514, footnote; *Electromagnetic Theory*, Vol. I, p. 273.)

PART II.

APPLICATIONS OF THE THEORY OF MOLECULAR POLARISATION OF MATERIAL MEDIA TO THE PHENOMENA OF ABERRATION AND CONVECTION.

19. WE shall follow Lorentz* in the introduction of a new variable t'', the "local time," to take the place of t'. The resulting theorem is more general than his transformation of § 59 inasmuch as it includes electric currents and media possessing magnetic susceptibility, as well as media with a specific inductive capacity.

We may commence with a problem in which material media are drifting with constant velocity $\mathbf{u}$, a function neither of the coordinates nor the time. The polarisation in the media being molecular, the appropriate equations will be those of § 14: referring to axes moving with the media, we find

$$\left.\begin{aligned} \frac{d\mathbf{E}}{dt'} + (K-1)\frac{d\mathbf{E}'}{dt'} + 4\pi\mathbf{C} &= V \operatorname{curl} \mathbf{H}' \\ \frac{d\mathbf{H}}{dt'} + (\mu - 1)\frac{d\mathbf{H}'}{dt'} &= -V \operatorname{curl} \mathbf{E}' \end{aligned}\right\}$$

or

$$\left.\begin{aligned} \frac{d\mathbf{D}}{dt'} + 4\pi\mathbf{C} &= V \operatorname{curl} \mathbf{H}' \\ \frac{d\mathbf{G}}{dt'} &= -V \operatorname{curl} \mathbf{E}' \end{aligned}\right\} \text{..............(42)},$$

with
$$\operatorname{div} \mathbf{D} = 4\pi\rho, \quad \operatorname{div} \mathbf{G} = 4\pi\tau \text{..............(43)}.$$

* *Versuch einer Theorie der electrischen und optischen Erscheinungen in bewegten Körpern* (Leyden, 1895). This work will in future be alluded to without explicit mention of its title.

At surfaces of discontinuity where μ, K may change suddenly, the vectors

$$(\mathbf{E}_2' - \mathbf{E}_1'),\ (\mathbf{H}_2' - \mathbf{H}_1') \text{ must be normal in direction} \quad \text{...(44)}.$$

We have also the dependent conditions that if $\mathbf{N}$ represent a unit length measured along the normal,

$$\left\{\mathbf{N}\left(\frac{d\mathbf{D}}{dt'} + 4\pi\mathbf{C}\right)\right\}_1^2 = 0, \quad \left\{\mathbf{N}\,\frac{d\mathbf{G}}{dt'}\right\}_1^2 = 0 \quad \text{......(45)}.$$

Further $\{\mathbf{ND}\}_1^2 = 4\pi\sigma$, $\{\mathbf{NG}\}_1^2 = 4\pi\upsilon$ .... (46),

where σ and υ are the electric and magnetic surface-densities.

Now let t' be replaced as an independent variable by the "local time" t'' given by

$$t'' = t' - (ux + vy + wz)/V^2,$$

and let x, y, z be then replaced by equal variables x'', y'', z''.

We shall have $\dfrac{d}{dt'} = \dfrac{d}{dt''}$, $\dfrac{d}{dx} = \dfrac{d}{dx''} - \dfrac{u}{V^2}\dfrac{d}{dt''}$, &c.,

i.e. $$\nabla = \nabla'' - \frac{\mathbf{u}}{V^2}\frac{d}{dt''},$$

and we shall distinguish all operators in the new system by two dashes. Thus

$$\begin{aligned}
\operatorname{div}\mathbf{D} &= \operatorname{div}''\mathbf{D} - \frac{1}{V^2}\left(\mathbf{u}\,\frac{d\mathbf{D}}{dt''}\right) \\
&= \operatorname{div}''\mathbf{D} - \frac{1}{V^2}\left(\mathbf{u}\,\frac{d\mathbf{D}}{dt'}\right) \\
&= \operatorname{div}''\mathbf{D} - \frac{1}{V}(\mathbf{u}\operatorname{curl}\mathbf{H}') + \frac{4\pi}{V^2}(\mathbf{uC}), \text{ by (42)} \\
&= \operatorname{div}''\mathbf{D} + \frac{1}{V}\operatorname{div}[\mathbf{uH}'] + \frac{4\pi}{V^2}(\mathbf{uC}) \quad \text{...... (47)},
\end{aligned}$$

and if squares of $\mathbf{u}/V$ are neglected,

$$\begin{aligned}
\operatorname{div}[\mathbf{uH}'] &= \operatorname{div}''[\mathbf{uH}] \\
&= V\operatorname{div}''[\mathbf{E}' - \mathbf{E}];
\end{aligned}$$

$\therefore$ by (47), $4\pi\rho = \operatorname{div}''(\mathbf{D} + \mathbf{E}' - \mathbf{E}) + \dfrac{4\pi}{V^2}(\mathbf{uC})$;

$$\therefore\ \operatorname{div}'' K\mathbf{E}' = 4\pi\{\rho - (\mathbf{uC})/V^2\} \quad \text{..........(48)}.$$

Similarly $$\text{div}''\,\mu\mathbf{H}' = 4\pi\tau \quad\text{.....................(49)}.$$

Now $$\text{curl}\,\mathbf{H}' = \text{curl}''\,\mathbf{H}' - \frac{1}{V^2}\left[\mathbf{u}\,\frac{d\mathbf{H}'}{dt''}\right],$$

$$\therefore\ \frac{d\mathbf{D}}{dt'} + 4\pi\mathbf{C} = V\,\text{curl}''\,\mathbf{H}' - \frac{1}{V}\frac{d}{dt'}[\mathbf{uH}],$$

$$\therefore\ \frac{d}{dt''}(K\mathbf{E}') + 4\pi\mathbf{C} = V\,\text{curl}''\,\mathbf{H}' \quad\text{.........(50)},$$

and, similarly,

$$\frac{d}{dt''}(\mu\mathbf{H}') = -\,V\,\text{curl}''\,\mathbf{E}' \quad\text{...............(51)}.$$

Also by Ohm's Law, if the conductivity is denoted by λ, $\mathbf{C} = \lambda\mathbf{E}'$.

The surface conditions are that the vectors $(\mathbf{E}_2' - \mathbf{E}_1')$, $(\mathbf{H}_2' - \mathbf{H}_1')$ shall be normal in direction: hence the normal components of $[\mathbf{u}(\mathbf{E}_2' - \mathbf{E}_1')]$, $[\mathbf{u}(\mathbf{H}_2' - \mathbf{H}_1')]$ must vanish.

But, neglecting squares, we have in each region,

$$K\mathbf{E}' = \mathbf{D} + \frac{1}{V}[\mathbf{uH}'],$$

$$\mu\mathbf{H}' = \mathbf{G} - \frac{1}{V}[\mathbf{uE}'],$$

and therefore

$$\{\mathbf{ND}\}_1^2 = \{\mathbf{N}K\mathbf{E}'\}_1^2, \quad \{\mathbf{NG}\}_1^2 = \{\mathbf{N}\mu\mathbf{H}'\}_1^2 \quad\text{........ (52)}.$$

These quantities are by (46) respectively equal to $4\pi\sigma$, $4\pi\upsilon$.

Further the vector $\mathbf{N}$ remains constant in magnitude and direction, and the equations (52) may be differentiated with respect to t' or to t'': on substituting in (45) we get

$$\left\{\mathbf{N}\left(\frac{dK\mathbf{E}'}{dt''} + 4\pi\mathbf{C}\right)\right\}_1^2 = 0, \quad \left\{\mathbf{N}\,\frac{d\mu\mathbf{H}'}{dt''}\right\}_1^2 = 0 \quad\text{...(53)}.$$

We shall now sum up the results:—

We commence with a system moving with uniform velocity $\mathbf{u}$. The equations of motion are

$$\left.\begin{aligned} \frac{d\mathbf{D}}{dt'} + 4\pi\mathbf{C} &= V\,\text{curl}\,\mathbf{H}' \\ \frac{d\mathbf{G}}{dt'} &= -\,V\,\text{curl}\,\mathbf{E}' \end{aligned}\right\} \quad\text{............(42)},$$

where $\mathbf{D} = \mathbf{E} + (K-1)\mathbf{E}'$, $\mathbf{G} = \mathbf{H} + (\mu - 1)\mathbf{H}'$,

$$\text{div}\,\mathbf{D} = 4\pi\rho, \quad \text{div}\,\mathbf{G} = 4\pi\tau \text{..............(43)}.$$

The boundary conditions are :—

That $\{\mathbf{E}'\}_1^2$, $\{\mathbf{H}'\}_1^2$, shall be normal .... (44),
with the dependent and superfluous equations :—

$$\left\{\mathbf{N}\left(\frac{d\mathbf{D}}{dt'} + 4\pi\mathbf{C}\right)\right\}_1^2 = 0, \quad \left\{\mathbf{N}\frac{d\mathbf{G}}{dt'}\right\}_1^2 = 0 \text{ ......(45)}.$$

Further $\{\mathbf{ND}\}_1^2 = 4\pi\sigma$, $\{\mathbf{NG}\}_1^2 = 4\pi\upsilon$ ....(46).

These equations are transformed by the introduction of new variables

$$x'' = x, \quad y'' = y, \quad z'' = z, \quad t'' = t' - (ux + vy + wz)/V^2,$$

and, when squares of $\mathbf{u}/V$ are neglected, the above system of equations becomes :—

$$\frac{d}{dt''}(K\mathbf{E}') + 4\pi\mathbf{C} = V\,\text{curl}''\,\mathbf{H}' \text{............(50)},$$

$$\frac{d}{dt''}(\mu\mathbf{H}') \qquad = - V\,\text{curl}''\,\mathbf{E}' \text{........(51)},$$

$$\text{div}''\,K\mathbf{E}' = 4\pi\{\rho - \mathbf{uC}/V^2\} \text{...........(48)},$$

$$\text{div}''\,\mu\mathbf{H}' = 4\pi\tau \text{ ....(49)},$$

with $\mathbf{C} = \lambda\mathbf{E}'$, as before.

The boundary conditions become :—

That $\{\mathbf{E}'\}_1^2$, $\{\mathbf{H}'\}_1^2$ shall be normal ....(44),
with the dependent equations

$$\left\{\mathbf{N}\left(\frac{d}{dt''}K\mathbf{E}' + 4\pi\mathbf{C}\right)\right\}_1^2 = 0, \quad \left\{\mathbf{N}\frac{d}{dt''}\mu\mathbf{H}'\right\}_1^2 = 0 \text{......(53)}.$$

Further from (46), (52),

$$\{\mathbf{N}K\mathbf{E}'\}_1^2 = 4\pi\sigma, \quad \{\mathbf{N}\mu\mathbf{H}'\}_1^2 = 4\pi\upsilon.$$

Consider now the same distribution of material media, referred to x'', y'', z'', t'', the media, the ether and the new axes being at rest.

If $\mathbf{E}'$, $\mathbf{H}'$ denote the electric and magnetic forces at a point at rest in the new system, then $\mathbf{C}\,(=\lambda\mathbf{E}')$ will be the new current and the above set of equations will be the ordinary equations of the field in the new system, except that, by (48), the new volume density of electricity (apart from that due to polarisation) will be $\rho - \mathbf{uC}/V^2$, where ρ was the density in the former case.

Hence corresponding to a distribution

$$\mathbf{E}, \mathbf{D}, \mathbf{H}, \mathbf{G}, \rho, \sigma, \tau, v, \mathbf{C},$$

in the case of drift, the time t' being referred to points moving with the material media, there will be a distribution

$$\mathbf{E}', K\mathbf{E}', \mathbf{H}', \mu\mathbf{H}', \rho - (\mathbf{uC})/V^2, \sigma, \tau, v, \mathbf{C},$$

which satisfy the conditions of a system at rest, the time t'' in this system being related to t' by the equation

$$t'' = t' - (ux + vy + wz)/V^2.$$

The material media involved are the same in all respects in the two cases.

Hence it follows (Lorentz, § 60), that the path of a ray of light remains a possible path of a ray, and that the ordinary laws of refraction and reflection still hold good when the media have a uniform motion.

From this theorem may be deduced all the ordinary facts connected with aberration. Of these one may be taken as an example: cf. Lorentz, § 68.

20. In Airy's "water-telescope" experiment the constant of aberration was determined by means of a telescope of which the tube could be filled with water. The value of the constant when deduced from observations made with the tube filled with water, was equal to its value as obtained with an ordinary telescope.

Let us take axes fixed in the ether, and consider the light coming from a star which lies in the direction l, m, n. The forces may be taken as proportional to $e^{is(Vt+lx+my+nz)}$, where $s = 2\pi/(\text{wave length})$.

If now the origin be taken at a point possessing the same velocity $\mathbf{u}$ as the earth, and x', y', z', t' be the coordinates and time referred to the new system, we shall have $t = t'$, and

$x = x' + ut'$, $y = y' + vt'$, $z = z' + wt'$: the forces are accordingly proportional to

$$e^{is(V't'+lx'+my'+nz')},$$

where $$V' = V + \Sigma ul \text{ ....(54).}$$

Now substitute t'', x'', y'', z'' given by

$$t'' = t' - \Sigma ux/V^2, \; x'' = x', \; y'' = y', \; z'' = z',$$

and the exponential factor becomes

$$e^{is(V't''+V'\Sigma ux/V^2+\Sigma lx'')} \equiv e^{is''(V''t''+\Sigma l''x'')} \text{ ....(55),}$$

where $$\frac{V'}{V''} = \frac{l + uV'/V^2}{l''} = \frac{m + vV'/V^2}{m''}$$

$$= \frac{n + wV'/V^2}{n''} = \frac{s''}{s} \text{ ....(56).}$$

Hence each fraction is equal to

$$(1 + 2V'\Sigma ul/V^2)^{\frac{1}{2}},$$

or $$1 + V'\Sigma ul/V^2,$$

or, by (54), $$1 + \Sigma ul/V,$$

neglecting squares. But, by (54), (56),

$$\frac{V'}{V} = 1 + \frac{\Sigma ul}{V} = \frac{V'}{V''};$$

$$\therefore \; V'' = V,$$

as we should expect, from the fact that (55) satisfies the ordinary electromagnetic equations for axes at rest.

Further, by (56)

$$\frac{l''}{l + u/V} = \frac{m''}{m + v/V} = \frac{n''}{n + w/V},$$

and by our transformation theorem we know that the ray (55) corresponds to the light in the actual experiment.

Hence if we consider the rays which come to a focus at a particular point in the retina of the observer's eye in the actual and in the transformed systems, we find that rays coming from a star in the direction (l, m, n) will appear to the observer to have come from a star which, if the earth were at rest, would lie in the direction (l'', m'', n'') given by

$$\frac{l''}{l+u/V}=\frac{m''}{m+v/V}=\frac{n''}{n+w/V}.$$

Thus the observed phenomena are explained: the direction of l'', m'', n'' is that of the resultant of $(Vl,\ Vm,\ Vn)$ and of (u, v, w): further, the period is quickened in the ratio of s'' to s, or of

$$\{V+(ul+vm+wn)\} \text{ to } V;$$

this is in accordance with Döppler's law.

Double-refraction of plane waves passing through a drifting medium.

21. We shall consider a dielectric moving with constant velocity u in a direction which we shall take as OX, the origin of reference being fixed in the ether. The polarisation in the medium will be treated as molecular and the ether as at rest. Plane waves are being propagated with velocity U in the direction l, m, n.

Let the components of force be given by

$$\frac{X}{\xi}=\frac{Y}{\eta}=\frac{Z}{\zeta}=\frac{L}{\lambda}=\frac{M}{\mu}=\frac{N}{\nu}=e^{is\,(Ut-lx-my-nz)}.$$

Then the electric forces at a point moving with the dielectric will, by (29), be

$$X'=\xi,\quad Y'=\eta-\nu u/V,\quad Z'=\zeta+\mu u/V,$$

where the exponentials are omitted;

$$\therefore\ \operatorname{div}\mathbf{E}'=-is\,\{l\xi+m\eta+n\zeta+(\mu n-m\nu)\,u/V\}.$$

The equations referred to the fixed axes will be {from (19), (28), (22) and (27)}:

$$\operatorname{div}\{\mathbf{E}+(K-1)\,\mathbf{E}'\}=4\pi\rho=0,\quad \operatorname{div}\mathbf{H}=4\pi\tau=0,$$

$$\frac{dX}{dt} + u \operatorname{div} \mathbf{E} + (K-1)\frac{dX'}{dt'} = V\left(\frac{dN}{dy} - \frac{dM}{dz}\right),$$

$$\frac{dY}{dt} + (K-1)\frac{dY'}{dt'} = V\left(\frac{dL}{dz} - \frac{dN}{dx}\right),$$

$$\frac{dZ}{dt} + (K-1)\frac{dZ'}{dt'} = V\left(\frac{dM}{dx} - \frac{dL}{dy}\right),$$

where

$$\frac{d}{dt'} = \frac{d}{dt} + u\frac{d}{dx}.$$

Also

$$\frac{dL}{dt} = V\left(\frac{dY}{dz} - \frac{dZ}{dy}\right),$$

$$\frac{dM}{dt} = V\left(\frac{dZ}{dx} - \frac{dX}{dz}\right),$$

$$\frac{dN}{dt} = V\left(\frac{dX}{dy} - \frac{dY}{dx}\right).$$

Hence we find

$$\left.\begin{aligned} &U\xi - u(l\xi + m\eta + n\zeta) + (K-1)(U - ul)\,\xi \\ &\qquad = V(n\mu - m\nu), \\ &U\eta \qquad + (K-1)(U-ul)(\eta - u\nu/V) \\ &\qquad = V(l\nu - n\lambda), \\ &U\zeta \qquad + (K-1)(U-ul)(\zeta + u\mu/V) \\ &\qquad = V(m\lambda - l\mu) \end{aligned}\right\} \text{...(57);}$$

$$\left.\begin{aligned} U\lambda &= V(m\zeta - n\eta) \\ U\mu &= V(n\xi - l\zeta) \\ U\nu &= V(l\eta - m\xi) \end{aligned}\right\} \text{......................(58).}$$

On multiplying the equations (57) by l, m, n and adding, we deduce

$$K\Sigma l\xi + (K-1)(\mu n - \nu m)\,u/V = 0 \text{............(59),}$$

which is equivalent to the statement that the electric volume-density is zero: it may be put into the form

$$\Sigma l\{X + (K-1)X'\} = 0,$$

or

$$l\mathfrak{X} + m\mathfrak{Y} + n\mathfrak{Z} = 0 \text{....................(60).}$$

From (58)

$$\left.\begin{aligned} l\lambda + m\mu + n\nu &= 0 \\ \lambda\xi + \mu\eta + \nu\zeta &= 0 \end{aligned}\right\} \quad \text{.....................(61).}$$

Also on multiplying the equations of (57) by λ, μ, ν and adding,

$$U\Sigma\,\lambda\xi - u\lambda\Sigma\,l\xi + (K-1)(U-ul)\,\Sigma\,\lambda\xi = 0.$$

Hence, by (61), $\qquad \lambda\Sigma\,l\xi = 0,$

and either $\lambda = 0$, or $\Sigma\,l\xi = 0$.

22. These cases must be discussed separately.

I. If $\lambda = 0$, the magnetic force in the ether, which by (61) lies in the plane of the wave-front, is perpendicular to the direction of drift.

II. If $\Sigma\,(l\xi) = 0$, then by (59)

$$\mu n - \nu m = 0;$$

therefore the plane which passes through OX, the direction of drift, and l, m, n will contain the line λ, μ, ν.

Hence the direction of the magnetic force is either perpendicular or parallel to the projection upon the wave-front of the direction of drift.

I. When $\lambda = 0$, we put (59) in the form

$$\frac{\Sigma l\xi}{(K-1)\,u/V} = \frac{m\nu - n\mu}{K},$$

and, multiplying the last two equations of (58) by n, $-m$, we have

$$U\,(n\mu - m\nu) = V\,\{\xi - l\Sigma\,l\xi\}.$$

Hence

$$\frac{\Sigma l\xi}{(K-1)\,u/V} = \frac{m\nu - n\mu}{K} = \frac{V\xi}{-KU + (K-1)\,ul}.$$

On substituting in the first equation of (57) we find, since $\Sigma l\xi$ does not vanish,

$$\{U + (K-1)(U-ul)\}\,\{-KU + (K-1)\,ul\} \\ - (K-1)\,u^2 + KV^2 = 0;$$

$$\therefore \ \{KU-(K-1)\,ul\}^2 = KV^2-(K-1)\,u^2,$$

and, if V' is defined by $KV'^2 = V^2$, we have

$$U = \frac{K-1}{K}\,ul \pm \left(V'^2 - \frac{K-1}{K^2}\,u^2\right)^{\frac{1}{2}}.$$

For the wave propagated in the positive direction, as far as squares of u/V,

$$U = V' + \frac{K-1}{K}\,ul - \frac{K-1}{2K^2}\,\frac{u^2}{V'} \quad \text{............(62).}$$

II. When $\Sigma\, l\xi = 0$, by (59), $\mu n - \nu m = 0$; hence by the first equation of (57) $\xi = 0$. By the remaining equations of (57), on substituting for λ, μ, ν from (58),

$$U\eta + (K-1)\,(U-ul)\left(\eta - \frac{u}{U}\,l\eta\right) = \frac{V^2}{U}\,\{\eta - m\Sigma l\xi\} = \frac{V^2}{U}\,\eta\,;$$

$$\therefore \ \{U^2 + (K-1)\,(U-ul)^2\}\,\eta = V^2\eta.$$

Similarly $\quad \{U^2 + (K-1)\,(U-ul)^2\}\,\zeta = V^2\zeta,$

and we have the result

$$KU^2 - 2\,(K-1)\,Uul + (K-1)\,u^2l^2 = KV'^2.$$

Therefore neglecting squares, for the positive wave,

$$U = V' + \frac{K-1}{K}\,ul - \frac{K-1}{2K^2}\,\frac{u^2l^2}{V'} \quad \text{............(63).}$$

23. Hence a uniform isotropic dielectric becomes double refracting when it moves through the ether; and, if θ be the angle between the direction of the drift and the normal to the wave-front, the two possible modes of wave propagation have a difference of velocity equal to $\dfrac{K-1}{2K^2}\,\dfrac{u^2\sin^2\theta}{V'}$. To the first approximation each obeys Fresnel's law of aberration.

It may be noticed that the relations between the two modes of propagation may be put in another manner.

The slower wave, I, (62), is characterised by $\lambda = 0$, i.e. $m\zeta = n\eta$; while in II, (63), we have $\xi = 0$.

Now in $(m\mathfrak{Z} - n\mathfrak{Y})$ the terms multiplied by the exponential $e^{is(Ut-lx-my-nz)}$ are

$$m\zeta - n\eta + (K-1)(m\zeta + m\mu u/V - n\eta + n\nu u/V),$$

or
$$K(m\zeta - n\eta) + (K-1)(m\mu + n\nu) u/V.$$

But by (61), when $\lambda = 0$, $m\mu + n\nu = 0$.

Therefore in I,

$$m\mathfrak{Z} - n\mathfrak{Y} = 0.$$

Also in general the latter equation of (61) may be put in the form $\lambda\mathfrak{X} + \mu\mathfrak{Y} + \nu\mathfrak{Z} = 0$, for

$$\lambda\{\xi + (K-1)\xi\} + \mu\{\eta + (K-1)(\eta - u\nu/V)\} + \nu\{\zeta + (K-1)(\zeta + u\mu/V)\}$$
$$= K(\lambda\xi + \mu\eta + \nu\zeta) = 0.$$

24. Collecting the results, we have

$$\Sigma l\mathfrak{X} = 0, \quad \Sigma lL = 0, \quad \Sigma\mathfrak{X}L = 0;$$

also $\mu = 1$ so that $\mathbf{H} = \mathbf{G}$.

Hence the directions of the complete electric and magnetic polarisations **D**, **G** must lie in the wave front at right angles to each other.

Further, there are only two modes in which propagation is possible. In Case I, corresponding to the slower velocity, the total electric polarisation is parallel to the projection of the direction of drift upon the wave front, in Case II, it is perpendicular to it. For the magnetic polarisation the words "parallel" and "perpendicular" must be interchanged.

Rotating Dielectric Plates.

25. When a charged disc is rapidly rotated, the convection currents due to the charges set up magnetic forces which have been observed and measured by Rowland and others. (*Ber. d. Berlin. Acad.*, 1876, p. 211; *Phil. Mag.* 27, p. 445, 1889.)

This result is in accordance with a theory, such as the present, which is based on the existence of convection currents.

Another experiment due to Röntgen (*Wied. Ann.* 35, p. 264, 1888; 40, p. 93, 1890) consists in rotating an uncharged glass disc between and parallel to the plates of a condenser. The existence of an external magnetic field was detected, but its intensity was not great enough for accurate measurement.

Röntgen tried a further experiment (pp. 267, 8). On the hypothesis that the ether has not the same velocity as the earth, it is flowing through the glass disc, and is there polarised if the condenser plates are charged. It might therefore, Röntgen argued, be expected that when the disc is not rotating, there would be a magnetic field due to the relative motion of the polarised ether. He found, however, that such a field, if existent, was too small to be detected.

We shall discuss this experiment on the hypothesis that the glass is molecularly polarised and the ether is at rest; we shall simplify the analysis without sacrificing any essential feature by considering the case of a spherical mass of dielectric rotating about an axis parallel to the lines of force of a uniform electric field.

26. Let the velocity $\mathbf{u}''$ of the earth relative to the ether be (A, B, C), and let the angular velocity of the sphere be n. If the axis of rotation be taken as OZ, and OX, OY preserve fixed directions, the velocity of any point x, y, z of the sphere will be $\mathbf{u} = (A - ny, B + nx, C)$.

The general equations referred to axes moving with velocity $\mathbf{u}''$, as stated in (20), p. 17, are represented by

$$\frac{dX}{dt''} + X\left(\frac{dv''}{dy} + \frac{dw''}{dz}\right) - Y\frac{du''}{dy} - Z\frac{du''}{dz} + (u - u'')\operatorname{div}\mathbf{E}$$

$$+ (K-1)\left\{\frac{dX'}{dt'} + X'\left(\frac{dv}{dy} + \frac{dw}{dz}\right) - Y'\frac{du}{dy} - Z'\frac{du}{dz}\right\}$$

$$+ 4\pi p = V\left(\frac{dN''}{dy} - \frac{dM''}{dz}\right),$$

and here, the conditions having become steady, $\frac{d\mathbf{E}}{dt''}$, $\frac{d\mathbf{E}'}{dt''}$, $\frac{d\mathbf{H}}{dt''}$, etc., are zero. Thus, since

$$\frac{d}{dt''} = \frac{d}{dt'} + (\mathbf{u}'' - \mathbf{u})\nabla,$$

$$\therefore \frac{d}{dt'} = -ny\frac{d}{dx} + nx\frac{d}{dy},$$

when operating on any of the forces.

The value of μ being unity, the equations which hold inside the sphere will be

$$\left.\begin{aligned} -ny \operatorname{div} \mathbf{E} + n(K-1)\left(-y\frac{dX'}{dx} + x\frac{dX'}{dy} + Y'\right) &= V\left(\frac{dN''}{dy} - \frac{dM''}{dz}\right) \\ nx \operatorname{div} \mathbf{E} + n(K-1)\left(-y\frac{dY'}{dx} + x\frac{dY'}{dy} - X'\right) &= V\left(\frac{dL''}{dz} - \frac{dN''}{dx}\right) \\ n(K-1)\left(-y\frac{dZ'}{dx} + x\frac{dZ'}{dy}\right) &= V\left(\frac{dM''}{dx} - \frac{dL''}{dy}\right) \end{aligned}\right\} \quad \ldots(64)$$

$$\left.\begin{aligned} -ny \operatorname{div} \mathbf{H} &= V\left(\frac{dY''}{dz} - \frac{dZ''}{dy}\right) \\ nx \operatorname{div} \mathbf{H} &= V\left(\frac{dZ''}{dx} - \frac{dX''}{dz}\right) \\ 0 &= V\left(\frac{dX''}{dy} - \frac{dY''}{dx}\right) \end{aligned}\right\} \quad \ldots(65).$$

For the space outside the sphere the equations are the same as inside, except that $K = 1$.

Now there is no distribution of permanent magnetism and div $\mathbf{H}$ accordingly vanishes. Hence, by (65), curl $\mathbf{E}''$ is zero. This reasoning holds for space inside and outside the sphere...(66).

Forces at points within the sphere will be distinguished by suffix 1 and those at outside points by suffix 2.

Let the intensity of the electric field due to the condenser be unity, and the radius of the sphere a. Then if A, B, C, n were zero, we should have

$$\mathbf{E}_1 = \nabla\left\{\frac{3}{K+2}z\right\}, \quad \mathbf{E}_2 = \nabla\left\{z - \frac{K-1}{K+2}\left(\frac{a}{r}\right)^3 z\right\},$$

and $$\mathbf{H}_1 = 0, \quad \mathbf{H}_2 = 0 \quad \text{.........................(67).}$$

In the case before us $\frac{u''}{V}, \frac{na}{V}$ are small quantities, and the values of $\mathbf{E}, \mathbf{H}$, will therefore differ only by small quantities from the values of (67) ....(68).

Also $$\mathbf{E}' = \mathbf{E} + \frac{1}{V}[\mathbf{uH}],$$

$$\mathbf{E}'' = \mathbf{E} + \frac{1}{V}[\mathbf{u''H}].$$

So that, if squares be neglected,

$$\mathbf{E}' = \mathbf{E}, \quad \mathbf{E}'' = \mathbf{E}.$$

Then the permanent electric volume-density being zero,

$$\text{div}\,\{\mathbf{E} + (K-1)\,\mathbf{E}'\} = 0,$$

$$\therefore \text{div}\,\mathbf{E} = 0 \quad \text{....................(69).}$$

Now since by (66) curl $\mathbf{E}'' = 0$, we may assume

$$\mathbf{E}_1 = \mathbf{E}_1'' = \nabla\left\{\frac{3}{K+2}z + \phi_1\right\},$$

$$\mathbf{E}_2 = \mathbf{E}_2'' = \nabla\left\{z - \frac{K-1}{K+2}\left(\frac{a}{r}\right)^3 z + \phi_2\right\},$$

where by (68) ϕ_1, ϕ_2 are small functions. Further by (69),

$$\text{div}\,\mathbf{E} = 0,$$

$$\therefore \nabla^2\phi_1 = 0, \quad \nabla^2\phi_2 = 0 \quad \text{................(70).}$$

The surface conditions are, by (39), (35), (36), that the tangential components of $\mathbf{E}'$, $\mathbf{H}'$ shall be continuous; and by (38), (41) that the difference between the normal components of total electric and of total magnetic polarisation, $\mathbf{D}$ and $\mathbf{G}$, on the two sides of an element of the surface shall each be constant; this constant must be zero, for the sphere is uncharged.

Now the values

$$\mathbf{E}_1 = \nabla \left\{ \frac{3}{K+2} z \right\},$$

$$\mathbf{E}_2 = \nabla \left\{ z - \frac{K-1}{K+2} \left(\frac{a}{r}\right)^3 z \right\},$$

satisfy all the surface conditions.

Hence ϕ_1, ϕ_2 must obey the following requirements:—

(1) Within the sphere $\nabla^2 \phi_1 = 0$.

(2) Outside the sphere $\nabla^2 \phi_2 = 0$.

(3) At the surface the tangential components of $\nabla \phi_1$, $\nabla \phi_2$ must be continuous, and also

$$K \frac{d\phi_1}{dr} = \frac{d\phi_2}{dr}.$$

Therefore ϕ_1, ϕ_2 are the potentials due to a dielectric sphere with no electric distribution at any point, and are constants which may be equated to zero.

Further, by the equations (64) for inside space, we obtain by substituting for $\mathbf{E}$, its value $\left(0,\ 0,\ \frac{3}{K+2}\right)$,

$$\text{curl } \mathbf{H}_1'' = 0.$$

For outside space we have $K = 1$ and div $\mathbf{E}_2 = 0$,

$$\therefore \text{ curl } \mathbf{H}_2'' = 0.$$

Thus $\mathbf{H}_1'' = \nabla \omega_1$, $\mathbf{H}_2'' = \nabla \omega_2$, where ω_1, ω_2 are small functions: also

$$\mathbf{H} = \mathbf{H}'' + \frac{1}{V} [\mathbf{u}'' \mathbf{E}],$$

$$\therefore \text{ div } \mathbf{H} = \text{div } \mathbf{H}'' + \frac{1}{V} (\mathbf{E} \text{ curl } \mathbf{u}'' - \mathbf{u}'' \text{ curl } \mathbf{E}),$$

$$\therefore\ 0 = \text{div } \mathbf{H}'',$$

$$\therefore\ \nabla^2 \omega_1 = 0, \quad \nabla^2 \omega_2 = 0 \quad \text{..............(71).}$$

Also

$$\left.\begin{aligned} L' &= L'' - \frac{nx}{V} Z \\ M' &= M'' - \frac{ny}{V} Z \\ N' &= N'' + \frac{nx}{V} X + \frac{ny}{V} Y \end{aligned}\right\},$$

$$\therefore \left.\begin{aligned} L_1' &= \frac{d\omega_1}{dx} - \frac{3}{K+2}\frac{nx}{V} \\ M_1' &= \frac{d\omega_1}{dy} - \frac{3}{K+2}\frac{ny}{V} \\ N_1' &= \frac{d\omega_1}{dz} \end{aligned}\right\},$$

and
$$\left.\begin{aligned} L_2' &= \frac{d\omega_2}{dx} - \frac{nx}{V}\left\{1 - \frac{K-1}{K+2}\frac{a^3}{r^5}(x^2+y^2-2z^2)\right\} \\ M_2' &= \frac{d\omega_2}{dy} - \frac{ny}{V}\left\{1 - \frac{K-1}{K+2}\frac{a^3}{r^5}(x^2+y^2-2z^2)\right\} \\ N_2' &= \frac{d\omega_2}{dy} + \frac{ny}{V}\left\{\frac{K-1}{K+2}\frac{a^3}{r^5}3yz\right\} + \frac{nx}{V}\left\{\frac{K-1}{K+2}\frac{a^3}{r^5}3xz\right\} \end{aligned}\right\}.$$

At the surface, the tangential component of $\mathbf{H}'$ is continuous, so that the vector $\mathbf{H}_2' - \mathbf{H}_1'$ must have the direction of the normal.

$$\therefore \frac{L_2' - L_1'}{x} = \frac{M_2' - M_1'}{y} = \frac{N_2' - N_1'}{z}.$$

$$\therefore \frac{\dfrac{d}{dx}(\omega_2 - \omega_1) - \dfrac{nx}{V}\dfrac{K-1}{K+2}\dfrac{3z^2}{a^2}}{x} = \frac{\dfrac{d}{dy}(\omega_2 - \omega_1) - \dfrac{ny}{V}\dfrac{K-1}{K+2}\dfrac{3z^2}{a^2}}{y}$$
$$= \frac{\dfrac{d}{dz}(\omega_2 - \omega_1) + \dfrac{nz}{V}\dfrac{K-1}{K+2}\dfrac{3(x^2+y^2)}{a^2}}{z}.$$

$$\therefore \frac{1}{x}\frac{d}{dx}(\omega_2 - \omega_1) = \frac{1}{y}\frac{d}{dy}(\omega_2 - \omega_1) = \frac{1}{z}\frac{d}{dz}(\omega_2 - \omega_1) + \frac{3n}{V}\frac{K-1}{K+2} \quad \text{..............(72).}$$

The normal magnetic polarisation is continuous,

$$\therefore \{xL + yM + zN\}_1^2 = 0,$$

i.e. the values of

$$xL'' + yM'' + zN'' + \frac{1}{V}\begin{vmatrix} x & y & z \\ A & B & C \\ X & Y & Z \end{vmatrix} \quad \text{......(73),}$$

must be continuous.

Now since the tangential component of **E** is continuous,

$$\frac{X_2 - X_1}{x} = \frac{Y_2 - Y_1}{y} = \frac{Z_2 - Z_1}{z},$$

$$\therefore \begin{vmatrix} x & y & z \\ A & B & C \\ X_2 - X_1 & Y_2 - Y_1 & Z_2 - Z_1 \end{vmatrix} = 0.$$

Accordingly the determinant in (73) is continuous,

$$\therefore\ xL'' + yM'' + zN'' \text{ must be continuous.}$$

Hence

$$x \frac{d}{dx}(\omega_2 - \omega_1) + y \frac{d}{dy}(\omega_2 - \omega_1) + z \frac{d}{dz}(\omega_2 - \omega_1) = 0,$$

i.e. $$\frac{d}{dr}(\omega_2 - \omega_1) = 0 \text{ at the surface} \qquad \text{(74)},$$

while ω_1, ω_2 satisfy in addition

$$\nabla^2 \omega_1 = 0, \quad \nabla^2 \omega_2 = 0 \text{ and (72).}$$

Hence $-\omega_1$, $-\omega_2$ are magnetic potentials which may be regarded as due to a certain distribution of currents in the surface; and if θ, ϕ be defined in the ordinary manner by

$$x = r \sin\theta \cos\phi, \quad y = r \sin\theta \sin\phi, \quad z = r\cos\theta,$$

it may be verified that the surface currents in the direction of θ, ϕ increasing are $-\dfrac{V}{4\pi a \sin\theta} \dfrac{d}{d\phi}(\omega_2 - \omega_1)$ and $\dfrac{V}{4\pi a} \dfrac{d}{d\theta}(\omega_2 - \omega_1)$.

From (72) we may deduce that

$$\frac{d}{d\phi}(\omega_2 - \omega_1) = 0,$$

$$\frac{d}{d\theta}(\omega_2 - \omega_1) = \frac{3na^2}{V} \frac{K-1}{K+2} \sin\theta \cos\theta;$$

thus the currents circulate in circles of latitude, the strength of flow being

$$\frac{3na}{4\pi} \frac{K-1}{K+2} \sin\theta\cos\theta.$$

Equating this to $-\dfrac{1}{a}\dfrac{d\Phi}{d\theta}$, where Φ is the current function, we find

$$\Phi = \frac{n}{8\pi} \frac{K-1}{K+2}(2z^2 - x^2 - y^2),$$

and it may be deduced that

$$\left.\begin{aligned}\omega_1 &= \frac{3n}{10V}\frac{K-1}{K+2}(2z^2-x^2-y^2)\\ \omega_2 &= \frac{n}{5V}\frac{K-1}{K+2}\left(\frac{a}{r}\right)^5(x^2+y^2-2z^2)\end{aligned}\right\}\quad\dots\dots\dots(75).$$

These values may also be obtained by direct analysis from the conditions (71), (72), (74).

27. The external magnetic force which would be directly given by experiment, is L'', M'', N'', the force at points fixed relative to the earth. We have found that this is derivable from a magnetic potential equal to

$$\frac{n}{5V}\frac{K-1}{K+2}\left(\frac{a}{r}\right)^5(2z^2-x^2-y^2);$$

and the forces accordingly vanish when the sphere has no angular velocity. The observations of Röntgen both when n is finite and zero are thus accounted for.

It will be noticed that while there is no permanent surface-density, there is an induced surface-density produced by the interior polarisation of the dielectric and equal to $\frac{3}{4V}\frac{K-1}{K+2}\frac{z}{a}$: the convection-current due to the rotation of this would be

$$\frac{3na}{4\pi}\frac{K-1}{K+2}\sin\theta\cos\theta.$$

The agreement between this and our former expression for the equivalent current might have been guessed, but does not appear to be *à priori* certain.

PART III.

STRESS IN AN ELECTROMAGNETIC FIELD.

28. WE shall in our first discussion assume with Hertz that the polarisation is entirely tubular, and that the ether and the matter have a common velocity **u** which varies from point to point.

The energy within the volume ω, contained in a small closed surface which moves at each point with velocity **u**, will be

$$\frac{1}{8\pi}\{K\mathbf{E}'^2 + \mu\mathbf{H}'^2\}\,\omega.$$

We write the forces with dashes because they are in action at points moving with the matter. The volume in question will, owing to the motion, be changing in dimensions, and the time-rate, at which the energy within it varies, bears to the volume ω a ratio (cf. Hertz, *Ges. Werke* II, pp. 278, 9)

$$\frac{1}{8\pi}\cdot\frac{1}{\omega}\left\{\omega\frac{d}{dt'}(K\mathbf{E}'^2 + \mu\mathbf{H}'^2) + (K\mathbf{E}'^2 + \mu\mathbf{H}'^2)\frac{d}{dt'}\omega\right\},$$

or $$\frac{1}{4\pi}\left\{K\mathbf{E}'\frac{d\mathbf{E}'}{dt'} + \mu\mathbf{H}'\frac{d\mathbf{H}'}{dt'}\right\} + \frac{1}{8\pi}\{K\mathbf{E}'^2 + \mu\mathbf{H}'^2\}\operatorname{div}\mathbf{u}.$$

Now, as in § 11, we have

$$K\frac{d\mathbf{E}'}{dt'} + K\mathbf{E}'\operatorname{div}\mathbf{u} - K\mathbf{E}'\nabla\,.\,\mathbf{u} + 4\pi\mathbf{C} = V\operatorname{curl}\mathbf{H}',$$

$$\mu\frac{d\mathbf{H}'}{dt'} + \mu\mathbf{H}'\operatorname{div}\mathbf{u} - \mu\mathbf{H}'\nabla\,.\,\mathbf{u} = -V\operatorname{curl}\mathbf{E}',$$

so that the ratio in question becomes

$$\frac{V}{4\pi}(\mathbf{E}' \operatorname{curl} \mathbf{H}' - \mathbf{H}' \operatorname{curl} \mathbf{E}') - \mathbf{E}'\mathbf{C}$$

$$+ \frac{1}{4\pi}(-K\mathbf{E}'^2 \operatorname{div} \mathbf{u} + K\mathbf{E}' . \mathbf{E}'\nabla . \mathbf{u} + \tfrac{1}{2}K\mathbf{E}'^2 \operatorname{div} \mathbf{u})$$

$$+ \frac{1}{4\pi}(-\mu\mathbf{H}'^2 \operatorname{div} \mathbf{u} + \mu\mathbf{H}' . \mathbf{H}'\nabla . \mathbf{u} + \tfrac{1}{2}\mu\mathbf{H}'^2 \operatorname{div} \mathbf{u}),$$

or, by VII, p. xiv, $\frac{V}{4\pi} \operatorname{div} [\mathbf{H}'\mathbf{E}'] - \mathbf{E}'\mathbf{C}$

$$+ \frac{1}{8\pi}\{2K\mathbf{E}' . \mathbf{E}'\nabla . \mathbf{u} - K\mathbf{E}'^2 \operatorname{div} \mathbf{u}$$

$$+ 2\mu\mathbf{H}' . \mathbf{H}'\nabla . \mathbf{u} - \mu\mathbf{H}'^2 \operatorname{div} \mathbf{u}\}.$$

The first term corresponds to the 'Poynting flux,' and the second gives the rate at which work is done in maintaining the conduction currents. The remaining terms may be put in the form

$$\frac{1}{8\pi}\Sigma u_x \{K(X'^2 - Y'^2 - Z'^2) + \mu(L'^2 - M'^2 - N'^2)\}$$

$$+ \frac{1}{4\pi}\Sigma(w_y + v_z)\{KY'Z' + \mu M'N'\},$$

where u_x, w_y stand for $\frac{du}{dx}$, $\frac{dw}{dy}$, &c.

These terms indicate that energy is expended at a rate which may naturally be interpreted as work done against a system of electro-magnetic stresses during the deformation of the volume in question. There will thus be due to the field, a set of tensions

$$\frac{1}{8\pi}\{K(X'^2 - Y'^2 - Z'^2) + \mu(L'^2 - M'^2 - N'^2)\},$$

and of shearing stresses

$$\frac{1}{4\pi}\{KY'Z' + \mu M'N'\}.$$

29. We shall now enquire as to the resultant force which the field would thus exert upon a small portion of the medium.

The force parallel to Ox due to the electric field only will be

$$\frac{K}{8\pi}\frac{d}{dx}(X'^2 - Y'^2 - Z'^2) + \frac{K}{4\pi}\left\{\frac{d}{dy}(X'Y') + \frac{d}{dz}(X'Z')\right\},$$

or

$$\frac{K}{4\pi}\left\{X'\left(\frac{dX'}{dx} + \frac{dY'}{dy} + \frac{dZ'}{dz}\right) + Y'\left(\frac{dX'}{dy} - \frac{dY'}{dx}\right) + Z'\left(\frac{dX'}{dz} - \frac{dZ'}{dx}\right)\right\}.$$

This gives, with the magnetic terms,

$$X'\rho + \frac{K\mu}{4\pi V}\left\{Y'\left(\frac{dN'}{dt'} + N' \operatorname{div} \mathbf{u} - \mathbf{H}'\nabla . w\right) - Z'\left(\frac{dM'}{dt'} + M' \operatorname{div} \mathbf{u} - \mathbf{H}'\nabla . v\right)\right\}$$
$$+ L'\tau + \frac{K\mu}{4\pi V}\left\{-M'\left(\frac{dZ'}{dt'} + Z' \operatorname{div} \mathbf{u} - \mathbf{E}'\nabla . w\right) + N'\left(\frac{dY'}{dt'} + Y' \operatorname{div} \mathbf{u} - \mathbf{E}'\nabla . v\right)\right\}$$
$$+ \frac{\mu}{V}\{N'q - M'r\},$$

or

$$X'\rho + L'\tau + \frac{\mu}{V}\{N'q - M'r\}$$
$$+ \frac{K\mu}{4\pi V}\left\{\frac{d}{dt'}(N'Y' - M'Z') + 2(N'Y' - M'Z') \operatorname{div} \mathbf{u} - Y'.\mathbf{H}'\nabla . w + Z'.\mathbf{H}'\nabla . v + M'.\mathbf{E}'\nabla . w - N'.\mathbf{E}'\nabla . v\right\}.$$

Now let $[\mathbf{E}'\mathbf{H}'] \equiv 4\pi V\{P', Q', R'\}$ and the force becomes

$$X'\rho + L'\tau + \frac{\mu}{V}\{N'q - M'r\}$$
$$+ K\mu\left\{\frac{dP'}{dt'} + 2P' \operatorname{div} \mathbf{u} + Q'v_x - P'v_y + R'w_x - P'w_z\right\},$$

i.e.

$$X'\rho + L'\tau + \frac{\mu}{V}(N'q - M'r)$$
$$+ K\mu\left\{\frac{dP'}{dt'} + P' \operatorname{div} \mathbf{u} + P'u_x + Q'v_x + R'w_x\right\}.$$

Expressed vectorially, the force is

$$\mathbf{E}'\rho + \mathbf{H}'\tau + \frac{\mu}{V}[\mathbf{CH}'] \\ + K\mu\left\{\frac{d\mathbf{P}'}{dt'} + \mathbf{P}' \operatorname{div} \mathbf{u} + \nabla_u (\mathbf{P}'\mathbf{u})\right\} \quad \ldots\ldots (76),$$

where $\mathbf{P}' \equiv (P', Q', R')$ and ∇_u indicates the operator ∇ applied to $\mathbf{u}$ only.

The last group of terms may by IX be put into the form

$$K\mu\left\{\frac{d\mathbf{P}'}{dt'} + \mathbf{P}' \operatorname{div} \mathbf{u} + \mathbf{P}'\nabla . \mathbf{u} + [\mathbf{P}' \operatorname{curl} \mathbf{u}]\right\} \ldots\ldots (77).$$

Hence in addition to the terms $\mathbf{E}'\rho$, $\mathbf{H}'\tau$, $\frac{1}{V}[\mathbf{CB}]$, the existence of which might have been anticipated, there will be a force which does not vanish in empty space.

This force, upon a volume ω unoccupied by matter, due to the stresses exerted upon it by the surrounding ether, is

$$\omega\left\{\frac{d\mathbf{P}'}{dt'} + \mathbf{P}' \operatorname{div} \mathbf{u} + \mathbf{P}'\nabla . \mathbf{u} + [\mathbf{P}' \operatorname{curl} \mathbf{u}]\right\},$$

reducing, when $\mathbf{u}$ is the same at all points, to $\frac{1}{4\pi V}\frac{d}{dt'}[\mathbf{E}'\mathbf{H}']$ per unit volume.

Even when the ether is at rest, this has the finite value

$$\frac{1}{4\pi V}\frac{d}{dt}[\mathbf{EH}] \quad \ldots\ldots\ldots\ldots\ldots\ldots\ldots (78).$$

30. Let us now investigate the values of the stresses involved in the hypothesis that the polarisation in the ether is "tubular," but that in a magnet and in a dielectric the polarisations are "molecular": the ether is still to be regarded as possessing the same velocity as the matter. The equations will now be

$$K\left(\frac{d\mathbf{E}'}{dt'} + \mathbf{E}' \operatorname{div} \mathbf{u} - \mathbf{E}'\nabla . \mathbf{u}\right) + 4\pi\mathbf{C} = V \operatorname{curl} \mathbf{H}',$$

and $$\mu\left(\frac{d\mathbf{H}'}{dt'} + \mathbf{H}' \operatorname{div} \mathbf{u} - \mathbf{H}'\nabla . \mathbf{u}\right) = - V \operatorname{curl} \mathbf{E}'.$$

These equations being the same as in § 28, the stresses will be identical with those just calculated.

Stress when the ether is at rest, and molecular polarisation moves through it.

31. In the succeeding analysis it will be assumed that the ether is fixed and that bodies, the particles of which are infinitely small, move through it. Since the ether is fixed there will be nothing unjustifiable in the supposition that, in the stress system exerted by the ether upon a small portion of matter, there may be forces of translation as well as couples.

There is, however, as we shall see later, one condition that should be satisfied. The expression for the resultant force which is in general exerted at any point between the ether and the material medium at that point must vanish where no matter is present, i.e. in free space.

32. As we have already found, if the expression for the localised energy of the field is given us, and we know the equations which determine the variations of the forces, then the stresses in the ether can be deduced. And inasmuch as the stresses and the energy are logically connected, it is undesirable, when both are unknown, to consider their values independently.

In Maxwell's theory, the justification of his expression for the energy must be the belief that it yields conclusions which agree with observation; in the present case, the expression for the energy being unknown, we must be prepared to proceed tentatively until consistent results are obtained.

Under one set of conditions, however, progress is easy. If **u** be constant and uniform, independent both of time and position, we shall have $\frac{\delta}{\delta t} = \frac{d}{dt}$, and the equations connecting the variations of the forces, when there are no conduction currents, reduce from (21) and (26), i.e. from

$$\left.\begin{aligned}\frac{\delta\mathbf{D}}{\delta t} &= V\,\mathrm{curl}\,\mathbf{H}'\\ \frac{\delta\mathbf{G}}{\delta t} &= -V\,\mathrm{curl}\,\mathbf{E}'\end{aligned}\right\},$$

where $\mathbf{D} = \mathbf{E} + (K-1)\,\mathbf{E}', \quad \mathbf{G} = \mathbf{H} + (\mu - 1)\,\mathbf{H}',$

to

$$\left.\begin{aligned}\frac{d\mathbf{D}}{dt'} &= V\,\mathrm{curl}\,\mathbf{H}'\\ \frac{d\mathbf{G}}{dt'} &= -V\,\mathrm{curl}\,\mathbf{E}'\end{aligned}\right\}.$$

When, therefore, the charges are constant and the forces are independent of the time,

$$\mathrm{curl}\,\mathbf{E}' = 0, \quad \mathrm{curl}\,\mathbf{H}' = 0:$$

therefore we may take

$$\mathbf{E}' = -\nabla\psi, \quad \mathbf{H}' = -\nabla\Omega.$$

Now if the charges of electricity and magnetism are represented by e, m respectively, and if these are all multiplied by n, the values of ψ, Ω will be multiplied by n. The work done in increasing n to $n + \delta n$ will be $\Sigma n\psi\,.\,e\delta n + \Sigma n\Omega\,.\,m\,\delta n$, for we may bring up $e\delta n$ at a rate which differs infinitesimally from $\mathbf{u}$ and so do work against the field equal to $-e\delta n\,.\int_\infty^P n\mathbf{E}'d\mathbf{s}$, where $\mathbf{E}'d\mathbf{s}$ is the scalar-product. This is exactly on the lines of Maxwell's determination of the potential energy, and the total work done in bringing up all the system is obtained by integrating to n from 0 to 1. The result is

$$\tfrac{1}{2}\Sigma e\psi + \tfrac{1}{2}\Sigma m\Omega$$

$$= \frac{1}{8\pi}\int dv\,\{\psi\,.\,\mathrm{div}\,\mathbf{D} + \Omega\,.\,\mathrm{div}\,\mathbf{G}\}$$

$$+ \frac{1}{8\pi}\int dS\,(\psi\,\{\mathbf{ND}\}_1^2 + \Omega\,\{\mathbf{NG}\}_1^2),$$

(where $\mathbf{N}$ is a line of unit length normal to S, a surface of discontinuity)

$$= -\frac{1}{8\pi}\int dv\,(\mathbf{D}\nabla\,.\,\psi + \mathbf{G}\nabla\,.\,\Omega),$$

by Green's Theorem. Hence the work done is

$$\frac{1}{8\pi}\int dv\,(\mathbf{DE}'+\mathbf{GH}').$$

Thus we are led to suppose that the energy is distributed through space and that the amount per unit volume is

$$\frac{1}{8\pi}(\mathbf{DE}'+\mathbf{GH}'),$$

or
$$\frac{1}{8\pi}\{\mathbf{EE}'+(K-1)\,\mathbf{E}'^2+\mathbf{HH}'+(\mu-1)\,\mathbf{H}'^2\}\quad\ldots(79).$$

33. When, however, $\mathbf{u}$ is variable, the determination breaks down, and a modification becomes necessary. As an example of the mode in which the expression for the energy of a field may require alteration, we may take the following:

If an ordinary magnetic field containing permanent and temporary magnetism, and conduction currents, be gradually and proportionally brought from infinity, the mechanical work done in getting the component portions into position may be shown to be, in Maxwell's notation*,

$$-\tfrac{1}{2}\int dv\,(A\alpha+B\beta+C\gamma)+\tfrac{1}{2}\int dv\,(uF+vG+wH),$$

or
$$-\tfrac{1}{2}\int dv\left\{\Sigma A\alpha+\frac{1}{4\pi}\Sigma a\alpha\right\},$$

where $A\,.\,B\,.\,C$ are components of permanent magnetism, and u, v, w here stand for the conduction current.

The work done in the meantime, by batteries or other means, in maintaining the constancy of the currents will (from Maxwell, § 580) be

$$\frac{1}{4\pi}\int dv\,\Sigma a\alpha.$$

Hence the total energy of the system is the sum

$$\frac{1}{8\pi}\int dv\,(\Sigma a\alpha-4\pi\Sigma A\alpha),$$

or
$$\frac{1}{8\pi}\int dv\,\mu\alpha^2,\ \text{for } a=\mu\alpha+4\pi A,\ \&c.$$

* *Electricity and Magnetism*, §§ 632, 634 (1892).

It will be seen that the mechanical work of bringing up the system from infinity, docs not, when currents are present as well as magnets, give us the complete expression for the energy; the production of a system of the more general character requires the expenditure of work by electromotive forces, i.e. by agency of a more general character.

In the present case, resuming our usual notation, we shall, for purposes of convenience, anticipate the evidence which makes it necessary to add $\mathbf{uP}$ to our expression (79) in order to get results consistent with the conservation of energy; here

$$\mathbf{P} \equiv \frac{1}{4\pi V}[\mathbf{EH}]$$

$$\therefore \text{ by II,} \qquad \mathbf{uP} = \frac{1}{8\pi V}\{\mathbf{E}[\mathbf{Hu}] + \mathbf{H}[\mathbf{uE}]\}$$

$$= \frac{1}{8\pi}\{\mathbf{E}(\mathbf{E}-\mathbf{E}') + \mathbf{H}(\mathbf{H}-\mathbf{H}')\};$$

and the function obtained by adding this to

$$\frac{1}{8\pi}\{\mathbf{DE}' + \mathbf{GH}'\},$$

is

$$\frac{1}{8\pi}\{\mathbf{E}^2 + (K-1)\mathbf{E}'^2 + \mathbf{H}^2 + (\mu-1)\mathbf{H}'^2\}.$$

Some discussion of the evidence for this expression will be found in § 38, after the consequent stresses have been deduced. But the consistent character of the results furnishes in itself a sufficient justification.

34. We now investigate the results of assuming that the energy W may be taken as

$$\frac{1}{8\pi}\{\mathbf{E}^2 + (K-1)\mathbf{E}'^2 + \mathbf{H}^2 + (\mu-1)\mathbf{H}'^2\}$$

per unit volume. We shall suppose that the media are isotropic, but not necessarily homogeneous, and that the magnetic polarisation has the same relation to the magnetic force as the electric polarisation to the electric force: cf. § 10, α.

The equations of the field, for a circuit moving with velocity $\mathbf{u}$, as determined in § 14, are

$$\left.\begin{aligned}&\frac{d\mathbf{D}}{dt'}+\mathbf{D}\operatorname{div}\mathbf{u}-\mathbf{D}\nabla\,.\,\mathbf{u}+4\pi\mathbf{C}=V\operatorname{curl}\mathbf{H}'\\&\frac{d\mathbf{G}}{dt'}+\mathbf{G}\operatorname{div}\mathbf{u}-\mathbf{G}\nabla\,.\,\mathbf{u}\qquad\qquad=-V\operatorname{curl}\mathbf{E}'\end{aligned}\right\}.$$

Thus

$$\begin{aligned}\frac{1}{\omega}\frac{d}{dt'}(W\omega)&=\frac{1}{4\pi}\left\{\mathbf{E}\frac{d\mathbf{E}}{dt'}+(K-1)\,\mathbf{E}'\frac{d\mathbf{E}'}{dt'}\right\}\\&\quad+\frac{1}{8\pi}\{\mathbf{E}^2+(K-1)\,\mathbf{E}'^2\}\operatorname{div}\mathbf{u}\\&\quad+\text{corresponding magnetic terms,}\\&=\frac{1}{4\pi}\left\{(\mathbf{E}-\mathbf{E}')\frac{d\mathbf{E}}{dt'}+\mathbf{E}'\frac{d\mathbf{D}}{dt'}\right\}\\&\quad+\frac{1}{8\pi}\{\mathbf{E}^2+(K-1)\,\mathbf{E}'^2\}\operatorname{div}\mathbf{u}\\&\quad+\text{corresponding magnetic terms,}\\&=\frac{1}{4\pi}\left\{(\mathbf{E}-\mathbf{E}')\frac{d\mathbf{E}}{dt'}+(\mathbf{H}-\mathbf{H}')\frac{d\mathbf{H}}{dt'}\right\}\\&\quad+\frac{1}{4\pi}\{\mathbf{E}'(-\mathbf{D}\operatorname{div}\mathbf{u}+\mathbf{D}\nabla\,.\,\mathbf{u}-4\pi\mathbf{C}+V\operatorname{curl}\mathbf{H}')\\&\qquad\qquad+\mathbf{H}'(-\mathbf{G}\operatorname{div}\mathbf{u}+\mathbf{G}\nabla\,.\,\mathbf{u}-V\operatorname{curl}\mathbf{E}')\}\\&\quad+\frac{1}{8\pi}\{\mathbf{E}^2+(K-1)\,\mathbf{E}'^2+\mathbf{H}^2+(\mu-1)\,\mathbf{H}'^2\}\operatorname{div}\mathbf{u},\\&=\frac{1}{4\pi V}\left\{[\mathbf{Hu}]\frac{d\mathbf{E}}{dt'}+[\mathbf{uE}]\frac{d\mathbf{H}}{dt'}\right\}-\mathbf{E}'\mathbf{C}\\&\quad+\frac{V}{4\pi}\operatorname{div}[\mathbf{H}'\mathbf{E}']\ \text{(by formula VII)}\\&\quad+\frac{1}{8\pi}\{\mathbf{E}^2-2\mathbf{DE}'+(K-1)\,\mathbf{E}'^2\\&\qquad\qquad+\mathbf{H}^2-2\mathbf{GH}'+(\mu-1)\,\mathbf{H}'^2\}\operatorname{div}\mathbf{u}\\&\quad+\frac{1}{4\pi}\{\mathbf{E}'.\,\mathbf{D}\nabla\,.\,\mathbf{u}+\mathbf{H}'.\,\mathbf{G}\nabla\,.\,\mathbf{u}\}\ \ldots\ldots\ldots\ldots(80),\\&=\frac{1}{4\pi V}\mathbf{u}\frac{d}{dt'}[\mathbf{EH}]-\mathbf{E}'\mathbf{C}+\frac{V}{4\pi}\operatorname{div}[\mathbf{H}'\mathbf{E}']\end{aligned}$$

$$+\frac{1}{8\pi}\Sigma u_x\{\mathbf{E}^2-2\mathbf{E}\mathbf{E}'-(K-1)\mathbf{E}'^2+2X'\mathfrak{X}$$

$$+\mathbf{H}^2-2\mathbf{H}\mathbf{H}'-(\mu-1)\mathbf{H}'^2+2L'\mathfrak{L}\}$$

$$+\frac{1}{4\pi}\Sigma u_y\{X'\mathfrak{Y}+L'\mathfrak{M}\}+\frac{1}{4\pi}\Sigma u_z\{X'\mathfrak{Z}+L'\mathfrak{N}\}.$$

Thus in addition to the Poynting Flux $\frac{V}{4\pi}[\mathbf{E}'\mathbf{H}']$, and the work done in maintaining the conduction currents, we have energy expended at a rate which indicates a force, due to the field, of strength $-\frac{1}{4\pi V}\frac{d}{dt'}[\mathbf{E}\mathbf{H}]$ or $-\frac{d\mathbf{P}}{dt'}$, and electromagnetic stresses against which work is performed at a rate

$$\frac{1}{8\pi}\Sigma u_x\{\mathbf{E}^2-2\mathbf{E}\mathbf{E}'-(K-1)\mathbf{E}'^2+2\mathfrak{X}X'\}$$

$$+\frac{1}{8\pi}\Sigma(w_y+v_z)(Y'\mathfrak{Z}+\mathfrak{Y}Z')$$

$$+\frac{1}{8\pi}\Sigma(w_y-v_z)(\mathfrak{Y}Z'-Y'\mathfrak{Z})$$

+ corresponding magnetic terms.

These electric terms give a tension parallel to OX of amount

$$\frac{1}{8\pi}\{\mathbf{E}^2-2YY'-2ZZ'+(K-1)(X'^2-Y'^2-Z'^2)\},$$

a shearing stress of amount

$$\frac{1}{8\pi}\{Y'\mathfrak{Z}+\mathfrak{Y}Z'\},$$

or

$$\frac{1}{8\pi}\{Y'Z+YZ'+2(K-1)Y'Z'\},$$

and a couple equal to

$$\frac{1}{4\pi}\{Y'Z-YZ'\} \quad\text{.....................(81).}$$

35. The resulting force Ξ per unit volume parallel to OX due to the field becomes (cf. Maxwell's Treatise, § 641, (13)),

$$-\frac{dP}{dt'} \text{ (occurring explicitly)}$$

$$+\frac{1}{8\pi}\frac{d}{dx}\{\mathbf{E}^2 - 2\mathbf{E}\mathbf{E}' - (K-1)\,\mathbf{E}'^2 + 2X'\mathfrak{X}\}$$

$$+\frac{1}{4\pi}\frac{d}{dy}\{X'\mathfrak{Y}\} + \frac{1}{4\pi}\frac{d}{dz}\{X'\mathfrak{Z}\}$$

+ the corresponding magnetic terms.

Therefore

$$4\pi\Xi = -4\pi\frac{dP}{dt'} + \mathbf{E}\mathbf{E}_x - \frac{d}{dx}(\mathbf{E}\mathbf{E}') - (K-1)\,\mathbf{E}'\mathbf{E}_x' - \tfrac{1}{2}K_x\mathbf{E}'^2$$

$$+\frac{d}{dx}(X'\mathfrak{X}) + \frac{d}{dy}(X'\mathfrak{Y}) + \frac{d}{dz}(X'\mathfrak{Z})$$

+ magnetic terms,

$$= -4\pi\frac{dP}{dt'} + (\mathbf{E}-\mathbf{E}')\,\mathbf{E}_x - \mathbf{E}\mathbf{E}_x' - (K-1)\,\mathbf{E}'\mathbf{E}_x' - \tfrac{1}{2}K_x\mathbf{E}'^2$$

$$+ X'\,\mathrm{div}\,\mathbf{D} + \mathbf{D}\nabla\,.\,X'$$

+ magnetic terms,

$$= -4\pi\frac{dP}{dt'}$$

$$+4\pi\rho X' + (\mathbf{E}-\mathbf{E}')\,\mathbf{E}_x - \mathbf{D}\mathbf{E}_x' + \mathbf{D}\nabla\,.\,X' - \tfrac{1}{2}K_x\mathbf{E}'^2$$

+ magnetic terms,

$$= -4\pi\frac{dP}{dt'}$$

$$+4\pi\rho X' + \frac{1}{V}[\mathbf{Hu}]\,\mathbf{E}_x + \mathfrak{Y}(X_y' - Y_x')$$

$$+\mathfrak{Z}(X_z' - Z_x') - \tfrac{1}{2}K_x\mathbf{E}'^2$$

$$+4\pi\tau L' + \frac{1}{V}[\mathbf{uE}]\,\mathbf{H}_x + \mathfrak{M}(L_y' - M_x')$$

$$+\mathfrak{N}(L_z' - N_x') - \tfrac{1}{2}\mu_x\mathbf{H}'^2.$$

Therefore

$$4\pi V\Xi = 4\pi V\left(-\frac{dP}{dt'} + \rho X' + \tau L'\right)$$

$$+\{[\mathbf{Hu}]\,\mathbf{E}_x+[\mathbf{uE}]\,\mathbf{H}_x\}-\tfrac{1}{2}V(K_x\mathbf{E}'^2+\mu_x\mathbf{H}'^2)$$

$$+\mathfrak{P}\frac{\delta\mathfrak{N}}{\delta t}-\mathfrak{Z}\frac{\delta\mathfrak{M}}{\delta t}$$

$$-\mathfrak{M}\left(\frac{\delta\mathfrak{Z}}{\delta t}+4\pi r\right)+\mathfrak{N}\left(\frac{\delta\mathfrak{P}}{\delta t}+4\pi q\right)$$

$$=4\pi V\left(-\frac{dP}{dt'}+\rho X'+\tau L'\right)+4\pi(q\mathfrak{N}-r\mathfrak{M})$$

$$+\{\mathbf{u}\,[\mathbf{E}_x\mathbf{H}]+\mathbf{u}\,[\mathbf{EH}_x]\}-\tfrac{1}{2}V(K_x\mathbf{E}'^2+\mu_x\mathbf{H}'^2)$$

$$+\mathfrak{P}\left(\frac{d\mathfrak{N}}{dt'}+\mathfrak{N}\operatorname{div}\mathbf{u}-\mathbf{G}\nabla\,.\,w\right)$$

$$-\mathfrak{Z}\left(\frac{d\mathfrak{M}}{dt'}+\mathfrak{M}\operatorname{div}\mathbf{u}-\mathbf{G}\nabla\,.\,v\right)$$

$$-\mathfrak{M}\left(\frac{d\mathfrak{Z}}{dt'}+\mathfrak{Z}\operatorname{div}\mathbf{u}-\mathbf{D}\nabla\,.\,w\right)$$

$$+\mathfrak{N}\left(\frac{d\mathfrak{P}}{dt'}+\mathfrak{P}\operatorname{div}\mathbf{u}-\mathbf{D}\nabla\,.\,v\right).$$

On putting $[\mathbf{DG}]/4\pi V\equiv(\mathfrak{P},\mathfrak{Q},\mathfrak{R})\equiv\mathfrak{Q}$, we obtain

$$\Xi=-\frac{dP}{dt'}+\rho X'+\tau L'+\frac{1}{V}(q\mathfrak{N}-r\mathfrak{M})$$

$$+\mathbf{uP}_x-(K_x\mathbf{E}'^2+\mu_x\mathbf{H}'^2)/8\pi$$

$$+\frac{d\mathfrak{P}}{dt'}+2\mathfrak{P}\operatorname{div}\mathbf{u}+v_x\mathfrak{Q}-v_y\mathfrak{P}+w_x\mathfrak{R}-w_z\mathfrak{P}$$

$$=\rho X'+\tau L'+\frac{1}{V}(q\mathfrak{N}-r\mathfrak{M})-(K_x\mathbf{E}'^2+\mu_x\mathbf{H}'^2)/8\pi$$

$$+\frac{d}{dt'}(\mathfrak{P}-P)+\mathfrak{P}\operatorname{div}\mathbf{u}+\mathbf{uP}_x+\mathfrak{Q}\mathbf{u}_x\ \ \ldots\ldots\ldots\ldots(82).$$

For the meaning of ρ, τ in this expression see (19), (28) in § 14.

36. The electromagnetic couple of (81) per unit volume, expressed vectorially, is

$$-\frac{1}{4\pi}\{[\mathbf{EE}']+[\mathbf{HH}']\},$$

or
$$-\frac{1}{4\pi V}\{[\mathbf{E}\,[\mathbf{uH}]]+[\mathbf{H}\,[\mathbf{Eu}]]\},$$

or
$$-\frac{1}{4\pi V}\{\mathbf{u}\,.\,\mathbf{HE}-\mathbf{H}\,.\,\mathbf{uE}+\mathbf{E}\,.\,\mathbf{uH}-\mathbf{u}\,.\,\mathbf{HE}\}.$$

By formula III this is $-\frac{1}{4\pi V}[\mathbf{u}[\mathbf{EH}]]$,

or
$$[\mathbf{Pu}] \quad\text{.........................(83).}$$

37. At a point at rest in free ether ρ, τ, $\mathbf{C}$, $\mathbf{u}$ vanish and $\mathfrak{Q}=\mathbf{P}$. Thus $\mathbf{F}$ is zero, as it should be.

When $\mathbf{u}=0$ and when the field is steady as well as stationary,

$$\Xi=\rho X+\tau L+\frac{1}{V}(q\mathfrak{R}-r\mathfrak{M})-\frac{1}{8\pi}(K_x\mathbf{E}^2+\mu_x\mathbf{H}^2),$$

i.e.
$$\mathbf{F}=\rho\mathbf{E}+\tau\mathbf{H}+\frac{1}{V}[\mathbf{CG}]-\frac{1}{8\pi}(\mathbf{E}^2\,.\,\nabla K+\mathbf{H}^2\,.\,\nabla\mu)\quad\text{...(84).}$$

In the case of homogeneous media, this is the expression which Maxwell's theory would lead us to expect.

38. Let us now examine the result of taking

$$W_1\equiv\frac{1}{8\pi}(\mathbf{DE}'+\mathbf{GH}')$$

as the potential energy. We find in that case, when the medium is homogeneous,

$$\frac{1}{\omega}\frac{d}{dt'}(W_1\omega)=\frac{1}{8\pi}\left\{\mathbf{E}\frac{d\mathbf{E}'}{dt'}+\frac{d\mathbf{E}}{dt'}\mathbf{E}'+2(K-1)\mathbf{E}'\frac{d\mathbf{E}'}{dt'}\right.$$
$$\left.+\mathbf{H}\frac{d\mathbf{H}'}{dt'}+\frac{d\mathbf{H}}{dt'}\mathbf{H}'+2(\mu-1)\mathbf{H}'\frac{d\mathbf{H}'}{dt'}\right\}$$
$$+\frac{1}{8\pi}(\mathbf{DE}'+\mathbf{GH}')\operatorname{div}\mathbf{u},$$
$$=\frac{V}{4\pi}\{\mathbf{E}'\operatorname{curl}\mathbf{H}'-\mathbf{H}'\operatorname{curl}\mathbf{E}'\}$$
$$+\frac{1}{4\pi}\{\mathbf{E}'(\mathbf{D}\nabla\,.\,\mathbf{u}-\mathbf{D}\operatorname{div}\mathbf{u}-4\pi\mathbf{C})+\mathbf{H}'(\mathbf{G}\nabla\,.\,\mathbf{u}-\mathbf{G}\operatorname{div}\mathbf{u})\}$$

$$+\frac{1}{8\pi}\left\{\mathbf{E}\frac{d\mathbf{E}'}{dt'}-\mathbf{E}'\frac{d\mathbf{E}}{dt'}+\mathbf{H}\frac{d\mathbf{H}'}{dt'}-\mathbf{H}'\frac{d\mathbf{H}}{dt'}\right\}$$

$$+\frac{1}{8\pi}\{\mathbf{DE}'+\mathbf{GH}'\}\operatorname{div}\mathbf{u},$$

$$=\frac{V}{4\pi}\operatorname{div}[\mathbf{H}'\mathbf{E}']-\mathbf{E}'\mathbf{C}$$

$$-\frac{1}{8\pi}\{\mathbf{DE}'+\mathbf{GH}'\}\operatorname{div}\mathbf{u}+\frac{1}{4\pi}\{\mathbf{E}'.\mathbf{D}\nabla.\mathbf{u}+\mathbf{H}'.\mathbf{G}\nabla.\mathbf{u}\}$$

$$+\frac{1}{8\pi V}\left\{\mathbf{E}\frac{d}{dt'}[\mathbf{uH}]-[\mathbf{uH}]\frac{d\mathbf{E}}{dt'}-\mathbf{H}\frac{d}{dt'}[\mathbf{uE}]+[\mathbf{uE}]\frac{d\mathbf{H}}{dt'}\right\}$$

$$=\frac{V}{4\pi}\operatorname{div}[\mathbf{H}'\mathbf{E}']-\mathbf{E}'\mathbf{C}$$

$$-\frac{1}{8\pi}\{\mathbf{DE}'+\mathbf{GH}'\}\operatorname{div}\mathbf{u}$$

$$+\frac{1}{4\pi}\Sigma u_x\{X'\mathfrak{X}+L'\mathfrak{L}\}$$

$$+\frac{1}{4\pi}\Sigma u_y\{X'\mathfrak{Y}+L'\mathfrak{M}\}+\frac{1}{4\pi}\Sigma u_z\{X'\mathfrak{Z}+L'\mathfrak{N}\}$$

$$+\frac{1}{8\pi V}\left\{\frac{d\mathbf{u}}{dt'}[\mathbf{HE}]+\mathbf{u}\left[\frac{d\mathbf{H}}{dt'}\mathbf{E}\right]-\mathbf{u}\left[\mathbf{H}\frac{d\mathbf{E}}{dt'}\right]\right.$$

$$\left.+\frac{d\mathbf{u}}{dt'}[\mathbf{HE}]+\mathbf{u}\left[\mathbf{H}\frac{d\mathbf{E}}{dt'}\right]-\mathbf{u}\left[\frac{d\mathbf{H}}{dt'}\mathbf{E}\right]\right\}\quad\ldots\ldots(85),$$

by use of formula II.

In this expression the first two terms correspond to the Poynting flux and the maintenance of the conduction currents. The last bracket reduces to

$$\frac{1}{4\pi V}\frac{d\mathbf{u}}{dt'}[\mathbf{HE}]\text{ or }-\mathbf{P}\frac{d\mathbf{u}}{dt'}\quad\ldots\ldots\ldots\ldots\ldots(86).$$

The remaining terms give stresses which are represented by:—

A tension parallel to OX of amount

$$\frac{1}{8\pi}\{-\mathbf{DE}'-\mathbf{GH}'+2X'\mathfrak{X}+2L'\mathfrak{L}\},$$

a shear

$$\frac{1}{8\pi}\{\mathfrak{Y}Z' + Y'\mathfrak{Z} + \mathfrak{M}N' + M'\mathfrak{N}\},$$

and a couple

$$\frac{1}{4\pi}\{\mathfrak{Y}Z' - Y'\mathfrak{Z} + \mathfrak{M}N' - M'\mathfrak{N}\}.$$

These stresses yield a force $\mathbf{F} \equiv (\Xi, \mathrm{H}, \mathrm{Z})$, per unit volume, given by

$$4\pi\Xi = -\tfrac{1}{2}\frac{d}{dx}(\mathbf{DE}' + \mathbf{GH}')$$

$$+\frac{d}{dx}(X'\mathfrak{X} + L'\mathfrak{L}) + \frac{d}{dy}(X'\mathfrak{Y} + L'\mathfrak{M}) + \frac{d}{dz}(X'\mathfrak{Z} + L'\mathfrak{N})$$

$$= -\tfrac{1}{2}\{\mathbf{E}_x\mathbf{E}' + \mathbf{E}\mathbf{E}_x' + 2(K-1)\mathbf{E}'\mathbf{E}_x'\} + X'\operatorname{div}\mathbf{D} + \mathbf{D}\nabla . X'$$

$+$ corresponding magnetic terms,

$$= \{\tfrac{1}{2}(\mathbf{E}\mathbf{E}_x' - \mathbf{E}_x\mathbf{E}' + \mathbf{H}\mathbf{H}_x' - \mathbf{H}_x\mathbf{H}') - (\mathbf{D}\mathbf{E}_x' + \mathbf{G}\mathbf{H}_x')\}$$

$$+ 4\pi\rho X' + 4\pi\tau L' + \mathbf{D}\nabla . X' + \mathbf{G}\nabla . L'$$

$$= \frac{1}{2V}\left\{\mathbf{E}\frac{d}{dx}[\mathbf{uH}] - \mathbf{E}_x[\mathbf{uH}] + \mathbf{H}\frac{d}{dx}[\mathbf{Eu}] - \mathbf{H}_x[\mathbf{Eu}]\right\}$$

$$+ 4\pi\rho X' + 4\pi\tau L' - \mathbf{D}\mathbf{E}_x' + \mathbf{D}\nabla . X' - \mathbf{G}\mathbf{H}_x' + \mathbf{G}\nabla . L'$$

$$= \frac{1}{2V}\{\mathbf{u}_x[\mathbf{HE}] + \mathbf{u}[\mathbf{H}_x\mathbf{E}] - \mathbf{u}[\mathbf{H}\mathbf{E}_x] + \mathbf{u}[\mathbf{H}\mathbf{E}_x]$$

$$+ \mathbf{u}_x[\mathbf{HE}] - \mathbf{u}[\mathbf{H}_x\mathbf{E}]\} + 4\pi\rho X' + 4\pi\tau L'$$

$$+ \mathfrak{Y}(X_y' - Y_x') + \mathfrak{Z}(X_z' - Z_x') + \mathfrak{M}(L_y' - M_x') + \mathfrak{N}(L_z' - N_x').$$

$$\therefore\ 4\pi V\Xi = \mathbf{u}_x[\mathbf{HE}] + 4\pi V(\rho X' + \tau L')$$

$$+ \mathfrak{Y}\frac{\delta\mathfrak{N}}{\delta t} - \mathfrak{Z}\frac{\delta\mathfrak{M}}{\delta t} - \mathfrak{M}\left(\frac{\delta\mathfrak{Z}}{\delta t} + 4\pi r\right) + \mathfrak{N}\left(\frac{\delta\mathfrak{Y}}{\delta t} + 4\pi q\right).$$

Hence, using the analysis of § 35, (82), we find

$$\Xi = \mathbf{u}_x\mathbf{P} + \rho X' + \tau L' + \frac{1}{V}(q\mathfrak{N} - r\mathfrak{M})$$

$$+ \frac{d\mathfrak{P}}{dt'} + \mathfrak{P}\operatorname{div}\mathbf{u} + \mathfrak{P}u_x + \mathfrak{Q}v_x + \mathfrak{R}w_x \quad\text{.........(87).}$$

If $K=1$, $\mu=1$ and $C=0$, then $(\mathfrak{P}, \mathfrak{Q}, \mathfrak{R}) = \mathbf{P}$, and

$$\Xi = \rho X' + \tau L' + \frac{dP}{dt'} + P \operatorname{div} \mathbf{u} + 2\mathbf{P}\mathbf{u}_x \text{ ....(88)},$$

$$\therefore\ \mathbf{F} = \rho\mathbf{E}' + \tau\mathbf{H}' + \frac{d\mathbf{P}}{dt'} + \mathbf{P} \operatorname{div} \mathbf{u} + 2\nabla_u . \mathbf{P}\mathbf{u} \text{...(89)}.$$

The hypothesis that the energy per unit volume is $(\mathbf{D}\mathbf{E}' + \mathbf{G}\mathbf{H}')/8\pi$ is thus subject to several objections. It yields in (86) a term $-\mathbf{P}\dfrac{d\mathbf{u}}{dt'}$, of which there is no physical explanation: and it gives in the expression (89) for $\mathbf{F}$ a term $\dfrac{d\mathbf{P}}{dt'}$, which does not vanish for a stationary element in free space.

Corresponding to this term in $\mathbf{F}$ there will be a time-rate of expenditure of energy against mechanical forces equal to

$$u\frac{dP}{dt'} + v\frac{dQ}{dt'} + w\frac{dR}{dt'} \text{ or } \mathbf{u}\frac{d\mathbf{P}}{dt'},$$

per unit volume.

Hence our assumption of W_1 as the energy has led analytically to the appearance of terms $-\mathbf{P}\dfrac{d\mathbf{u}}{dt'}$, $-\mathbf{u}\dfrac{d\mathbf{P}}{dt'}$ in the rate of increase of the electromagnetic energy of the field; and these terms are, as far as we are aware, entirely unjustified by experience.

We are thus led to try the effect of adding to W_1 a term $\mathbf{u}\mathbf{P}$: the result is the expression for W investigated in §§ 34—37, and the consequent removal of the difficulties.

39. Let us consider a disturbance which is spreading through a region void of matter with a velocity equal to that of light. If we imagine the moving surface which bounds the disturbance to be, for the moment, material, and if the axis OZ be taken in the direction of the normal to the surface, the positive side of XOY being that unreached by the disturbance, then on this side $\mathbf{E}'$, $\mathbf{H}'$ are zero. But the tangential components of $\mathbf{E}'$, $\mathbf{H}'$ are continuous. Hence on the negative side

X', Y', L', M' are zero. Now the tension of the negative side parallel to ZO is, by (81),

$$\frac{1}{8\pi}\{\mathbf{E}^2 - 2XX' - 2YY' + \mathbf{H}^2 - 2LL' - 2MM'\},$$

and that on the positive side is zero.

Thus the motion of the surface, regarded as material, would be retarded by a force per unit area equal to

$$\frac{1}{8\pi}\{\mathbf{E}^2 + \mathbf{H}^2\},$$

which is the potential energy per unit volume. This is as it should be; for the rate at which work is involved in the motion of the surface must be equal to the rate at which energy spreads through the medium.

40. The stresses (81) which we have determined from consideration of the energy of the ether will be stresses in the ether, and may naturally be regarded as the mechanism by which the ponderomotive forces, which act on charged bodies and on current circuits, are produced.

The resultant force due to the stresses upon an element of volume of a medium which is moving through the ether may, in view of the previous section, be expected to contain terms which are due to the motion and are independent of the polarisation at the point in question: but the force on an element at rest in the ether should contain nothing which does not vanish when there is no material medium there.

In all cases of the motion of finite bodies with which we have to deal, $\mathbf{u}/V$ is comparable with 10^{-4}, so that the numerical error due to neglecting the ratio will not be important: we shall accordingly discuss in greater detail the case in which $\mathbf{u} = 0$.

41. We then have a force $-\frac{d\mathbf{P}}{dt}$ occurring explicitly, a tension parallel to Ox equal to

$$\frac{1}{8\pi}\{K(X^2 - Y^2 - Z^2) + \mu(L^2 - M^2 - N^2)\},$$

and a shearing stress

$$\frac{1}{4\pi}(KYZ + \mu MN).$$

The couple vanishes ....(90).

By taking Ox along the direction of the resultant electric force, we realise that the above electric stresses *in the ether* are equivalent to a tension of $\frac{K}{8\pi}\mathbf{E}^2$ along lines of electric force, and a pressure $\frac{K}{8\pi}\mathbf{E}^2$ at right angles to them. This agrees with Maxwell's specification.

As long as we adopt the hypothesis that matter drags the ether with it, or that the ether and material bodies always have exactly the same motion, it is difficult to distinguish between stresses in the ether and stresses in the body: any displacement would be opposed by each of these stresses. It is accordingly a widespread if not a universal habit to regard the stresses as exerted by the material media: cf. Helmholtz, Kirchhoff and most writers on electro-striction and magneto-striction. The difficulties that manifest themselves are serious and well known. Maxwell's stresses give a resultant force whose component along OX is equal to

$$X\rho + L\tau + \frac{\mu}{V}(qN - rM) + K\mu\frac{dP}{dt} \quad \text{.........(91)},$$

and the last term of this expression does not vanish in empty space. Further it is not easy to see how a uniform, absolutely continuous, and homogeneous liquid medium can be in equilibrium if it be subject to the shearing-stresses set up according to Maxwell's theory by an electromagnetic field. Apart from these difficulties, it is customary to explain the forces between two charges in free space as due to stresses in the *ether*, while in a dielectric, such as glass, the Maxwellian stresses are supposed to reside in the material medium: and it does not appear logical to suppose that the stresses are exerted by the ether alone in the absence of bodies, and by material media when they are present.

When however the possibility of motion of a medium

through the ether is allowed, the difficulties may, I think, be overcome.

The resultant force is now, u/V being still neglected, given by

$$\mathbf{F} = \rho\mathbf{E} + \tau\mathbf{H} + \frac{\mu}{V}[\mathbf{CH}] + (\mu K - 1)\frac{d\mathbf{P}}{dt} \text{......(92)},$$

and the first difficulty does not occur. Further, the force on an element of a non-conductor in an electrostatic field is $\rho\omega\mathbf{E}$, where ω is the volume of the element. Thus $e\mathbf{E}$ will be the force exerted by the ether upon a particle whose charge is e.

42. According to the present theory, a body or material medium is to be regarded as an assemblage of infinitely small particles charged with electricity $\pm e$ or magnetism $\pm m$. Upon these the field exerts, by means of the stresses in the *ether*, in which the particles may be regarded as imbedded, forces $\pm \mathbf{E}e$, $\pm \mathbf{H}m$: the stresses (90) are the mean stresses over a small volume containing a large number of these particles. The stresses which are exerted within the *bodies* will be those required by the ordinary laws of elasticity in order that the equilibrium may be maintained in the presence of these forces $\pm \mathbf{E}e$, $\pm \mathbf{H}m$ as well as the externally applied mechanical stresses.

Some light may perhaps be thrown on this way of distinguishing stresses in the ether from stresses in a body, by the consideration of what happens in the case of gravitational attraction. If we interpret the action of gravity as due to stresses in the ether surrounding the infinitely small particles of a body, then the resultant effect of these stresses on a particle of mass $\mathfrak{m}$ will be $\mathfrak{m}\mathbf{R}$, where $\mathbf{R}$ is the force at the point. And the stresses *in the material body* will be those which are given by the ordinary elastic-solid theories as to the behaviour of the body under the action of a force equal to $\mathbf{R}$ per unit mass. The stresses in the body depend on the quantity $\mathbf{R}$ and on the external forces, such as pressures, applied to the body; but *the body-stresses are distinct from the stresses in the ether to which the forces* $\mathfrak{m}\mathbf{R}$ *are due.*

In the case of a polarised non-conducting dielectric, the

stresses exerted by the ether on the charged particles of the dielectric give, when the field is steady, the force $\rho\mathbf{E}$ per unit volume. When, as is usually the case, the dielectric possesses no volume-density, $\rho = 0$, and the force vanishes. The effect of an electric field upon the mechanical stresses in such a medium will be limited to the influence of forces at the boundary. Thus the internal equilibrium of a fluid dielectric presents no difficulties.

Owing to the linearity of the equations of elasticity, the stresses in the material media will obviously be the sums of the stresses due to the surface-forces and to the forces in the interior of the media.

43. At the surface of discontinuity between two media the stresses in the ether will give rise to a force per unit area which may be determined in a manner similar to that already employed. If the origin be taken in the surface and if the normal drawn into the second medium be taken as OX, then there will be stresses in the ether of the second medium which produce, on an element of area α of the bounding surface, component forces parallel to the axes

$$\alpha K_2 (X_2^2 - Y_2^2 - Z_2^2)/8\pi, \quad \alpha K_2 X_2 Y_2/4\pi, \quad \alpha K_2 X_2 Z_2/4\pi:$$

those on the negative side will be

$$-\alpha K_1 (X_1^2 - Y_1^2 - Z_1^2)/8\pi, \quad -\alpha K_1 X_1 Y_1/4\pi, \quad -\alpha K_1 X_1 Z_1/4\pi.$$

Also $\quad \{KX\}_1^2 = 4\pi\sigma, \quad \{Y\}_1^2 = 0, \quad \{Z\}_1^2 = 0.$

Thus the resultant force per unit area has components

$$\{K_2 X_2^2 - K_1 X_1^2 + (K_1 - K_2)(Y^2 + Z^2)\}/8\pi, \ \sigma Y, \ \sigma Z \ldots\ldots(93).$$

If the surface was uncharged before the polarisation was brought about, $\sigma = 0$, and the force due to the ether-stresses is normal.

When, in the general case, the direction cosines of the normal drawn into the second region are (l, m, n), the force per unit area of an uncharged surface will be

$$\{2K(lX + mY + nZ)^2 - K(X^2 + Y^2 + Z^2)\}_1^2/8\pi \ldots\ldots(94).$$

Verification of the previous section.

44. Inasmuch as the force per unit volume of a homogeneous medium is $\rho\mathbf{E}$, it might appear that when a surface of discontinuity is uncharged there could be no resultant force on it. Our analysis of §§ 34—37, however, provides the explanation of the difficulty, and we shall investigate the surface-forces by a second method.

Let us suppose that instead of a surface of discontinuity between two media there is a region of continuous transition and that, although some of the quantities vary with extreme rapidity as we pass from one medium to the other, the equations of the field still hold within this region of transition. Let the origin be taken at any point in the bounding surface and the axis OX along the normal into the second medium.

When K is a function of the coordinates, in an electrostatic field, by (84)

$$\Xi = \rho X - K_x\mathbf{E}^2/8\pi.$$

Hence, integrating through the region of transition, we have in the immediate neighbourhood of the origin,

$$8\pi \int \Xi \, dx = \int dx \, (8\pi\rho X - K_x\mathbf{E}^2)$$

$$= \int dx \, \{2X \operatorname{div} K\mathbf{E} + 2K\mathbf{E}\nabla . X - 2K . \mathbf{E}\mathbf{E}_x - K_x\mathbf{E}^2\}$$

(for when the field is electrostatic, curl $\mathbf{E} = 0$, and $\mathbf{E}\nabla . X = \mathbf{E}\mathbf{E}_x$)

$$= \int dx \left\{ 2\frac{d}{dx}(X . KX) + 2\frac{d}{dy}(X . KY) + 2\frac{d}{dz}(X . KZ) - \frac{d}{dx}(K\mathbf{E}^2)\right\}$$

$$= \int dx \left(\frac{d}{dx}\{K(X^2 - Y^2 - Z^2)\} + 2\frac{d}{dy}\{KXY\} + 2\frac{d}{dz}\{KXZ\}\right).$$

Now the only differential coefficients which are large within the region of integration are those with respect to x: hence, since the path of integration with respect to x is indefinitely small, there results in the limit,

$$\int \Xi dx = \{K(X^2 - Y^2 - Z^2)/8\pi\}_1^2,$$

which is our former result.

Similarly

$$\int \mathrm{H} dx = \frac{1}{8\pi}\int dx\left(2\frac{d}{dx}\{KXY\} + \frac{d}{dy}\{K(Y^2 - X^2 - Z^2)\} + 2\frac{d}{dz}\{KYZ\}\right)$$

$$= \{KXY/4\pi\}_1^2$$

$$= \sigma Y, \text{ as before : and so for } \int Z dz.$$

If one of the media in contact be a solid, its rigidity will provide the force necessary to overcome the surface forces just given. When both media are fluids and one is a liquid, the surface being initially uncharged, then the purely normal force due to the electric stresses is capable of equilibration through the agency of surface tension and the hydrostatic pressure of the fluids. The configuration in which the media can rest is given analytically by the representative equation

$$\{K_2X_2^2 - K_1X_1^2 + (K_1 - K_2)(Y^2 + Z^2)\}/8\pi + \Pi_1 - \Pi_2 + T(1/\rho_1 + 1/\rho_2) = 0 \ldots\ldots (95),$$

where Π_1, Π_2 are the pressures in the fluids, T is the surface tension and ρ_1, ρ_2 the radii of curvature of the bounding surface.

In a magnetic field similar results should, *on the present hypothesis*, be expected.

45. Our analysis has from § 11 proceeded on the assumption that the relation between magnetic polarisation and magnetic force is identical with that between electric polarisation and electric force. Certain facts suggest however that the similarity of relationship is open to question, and that a fundamental difference may exist between the two kinds of polarisation. Thus it is easy to obtain a body in which electric charge

of one sign preponderates; a corresponding excess of magnetic charge is unknown. In electricity conduction and decomposition by electrolysis are familiar phenomena; in magnetism they do not occur. According to one of the simplest modes of interpreting the phenomena of an electric field electric polarisation is explained in terms of a large number of electric charges of opposite signs. We have already seen that the quantitative relations deduced from the hypothesis of 'doublet' polarisation in an electric field agree well with the results of experimental investigation.

The term 'electron' has been used to designate points which have an electric charge and are without mass: but the present analysis does not depend on assumptions as to whether mass is or is not associated with the electric charges, and we shall therefore choose a different name and speak of 'ions.'

It is true that the term 'ion' has usually been applied to the electrified particles which are in motion during electrolysis: but we may perhaps be allowed to give the word a wider significance. We shall accordingly define an 'ion' as one of the aggregate of electrically charged particles which together constitute that portion of a material medium which responds to electric force.

The ions in an electrolyte or a conductor appear to be capable of individual motion. In a magnetic field however the smallest unit which is required for the expression of observed facts is a 'doublet'; in this we have two 'poles' equal in magnitude and opposite in sign.

Now an appropriate assemblage of ions describing small closed orbits would produce the same magnetic force as a distribution of magnetic doublets. And such an assemblage would have all the other characteristics of magnetisation which we have just noted: there would be no possibility of an excess of magnetism of one sign.

46. In the first place it may be of service to form a rough mental picture to illustrate the kind of way in which we may interpret the laws governing the ratios of electric and magnetic polarisation to electric and magnetic force.

We may regard a molecule as consisting of an assemblage of ions (comparable in number perhaps with a thousand) which are all describing small paths under the influence of their mutual electric forces of attraction or repulsion and of the forces due to the field: the electric force $\mathbf{E}''$ acting on an ion whose velocity is $\mathbf{u}''$ will be modified by the motion in accordance with equation (29), p. 18.

The numbers of positive and of negative ions in a molecule being the same we may for analytical purposes regard them as grouped in pairs to form doublets. The intensity $(\mathfrak{X}', \mathfrak{Y}', \mathfrak{Z}')$ of polarisation may then be determined in the following manner.

Let the charge on an ion be denoted by e and let its coordinates be denoted by (x, y, z). Then, if the summation Σex be effected over all ions lying within a small volume ω containing a large number of molecules, we shall have $\Sigma ex = \mathfrak{X}'\omega$ and similarly $\Sigma ey = \mathfrak{Y}'\omega$, $\Sigma ez = \mathfrak{Z}'\omega$. The intensity so defined is obviously independent of the origin of coordinates and of the arbitrary manner in which the grouping in doublets is brought about.

When no external electric force disturbs the motion of the ions, then, over a small region containing a large number of them, the moment per unit volume will be zero. When however a field of electric force represented by $\mathbf{E}e^{ipt}$ is exerted, we should expect the urging of the positive ions in one direction and the negative in the other to yield a polarisation $(\alpha + i\beta)\mathbf{E}e^{ipt}$ where α, β are real functions of p. Thus dispersion effects would be brought about; when the force is steady, $p = 0$ and $(\alpha + i\beta)$ becomes real: it is denoted by $(K-1)/4\pi$.

47. Magnetisation must according to this interpretation be explained by supposing that in a steady magnetic field the distribution of motion of the ions in a molecule is of such a character that the magnetic particle equivalent to the molecule has a moment proportional to the strength of the magnetic force. When the magnetic field is alternating and has an extremely short period, short by comparison with the time in which an ion describes its orbit, we should expect that the orbital motions would on the average remain unaffected, and

hence that the magnetic susceptibility would be zero. Such is apparently the case when light waves fall on a magnetic body*.

Let us now consider the effect of the distortion of a medium which is in a steady magnetic field. The molecules will change in their relative positions: and, in our present ignorance as to the configuration of a molecule and the laws which govern its equivalent magnetic moment, either of the hypotheses of § 10 is conceivable. Experimental determinations of the effects of magneto-striction in iron, nickel and cobalt indicate that in them the relation between the magnetic constants and the state of strain is more complex.

According to the former law (α) the ratio of the magnetic moment per unit volume remains proportional to the magnetic force. The consequences have been worked out in §§ 34—37.

According to (β), which appears to be the most natural hypothesis in the case of a fluid medium, the moment of the magnetic doublet equivalent to a molecule bears a constant ratio to the magnetic force in action upon it. And when the molecules become more closely packed, the susceptibility (as defined in the ordinary way and dependent on the ratio of the moment per unit volume to the magnetic force) will increase.

The results of this assumption will now be investigated.

48. We have seen in § 9 that if the magnetic moments of the individual molecules remain constant while their centres move with velocity $\mathbf{u}$, then 4π times the flux through a circuit moving with velocity $\mathbf{u}''$ will be the normal component of

$$(\mathbf{u}''-\mathbf{u})\operatorname{div}\mathbf{G}'-\mathbf{G}'\nabla.\mathbf{u} \quad\text{..............(5)},$$

or

$$(\mu-1)\{(\mathbf{u}''-\mathbf{u})\operatorname{div}\mathbf{H}'-\mathbf{H}'\nabla.\mathbf{u}\}.$$

Hence if the magnetic force $\mathbf{H}'$ varies and the moments of the separate molecules are proportional to $\mathbf{H}'$, the flux will be

$$\frac{\mu-1}{4\pi}\left\{\frac{d\mathbf{H}'}{dt'}-\mathbf{H}'\nabla.\mathbf{u}+(\mathbf{u}''-\mathbf{u})\operatorname{div}\mathbf{H}'\right\}.$$

* See Thomson's *Recent Researches*, § 357.

Thus the second equation of the field becomes

$$\frac{d\mathbf{H}}{dt''} + \mathbf{H}\operatorname{div}\mathbf{u}'' - \mathbf{H}\nabla\,.\,\mathbf{u}'' + (\mu - 1)\left\{\frac{d\mathbf{H}'}{dt'} - \mathbf{H}'\nabla\,.\,\mathbf{u} + (\mathbf{u}'' - \mathbf{u})\operatorname{div}\mathbf{H}'\right\} + 4\pi(\mathbf{u} - \mathbf{u}'')\tau = -V\operatorname{curl}\mathbf{E}''.$$

This reduces, by means of $\operatorname{div}\{\mathbf{H} + (\mu - 1)\mathbf{H}'\} = 4\pi\tau$, to

$$\frac{d\mathbf{H}}{dt''} + \mathbf{H}\operatorname{div}\mathbf{u}'' - \mathbf{H}\nabla\,.\,\mathbf{u}'' + (\mathbf{u} - \mathbf{u}'')\operatorname{div}\mathbf{H} + (\mu - 1)\left\{\frac{d\mathbf{H}'}{dt'} - \mathbf{H}'\nabla\,.\,\mathbf{u}\right\} = -V\operatorname{curl}\mathbf{E}'' \quad \text{......(96).}$$

We have, in addition, the former equation (20),

$$\frac{d\mathbf{E}}{dt''} + \mathbf{E}\operatorname{div}\mathbf{u}'' - \mathbf{E}\nabla\,.\,\mathbf{u}'' + (\mathbf{u} - \mathbf{u}'')\operatorname{div}\mathbf{E} + (K - 1)\left\{\frac{d\mathbf{E}'}{dt'} + \mathbf{E}'\operatorname{div}\mathbf{u} - \mathbf{E}'\nabla\,.\,\mathbf{u}\right\} + 4\pi\mathbf{C} = V\operatorname{curl}\mathbf{H}'' \quad \text{...........(20).}$$

Taking $\mathbf{u}''$ as $\mathbf{u}$ and simplifying, the equations become

$$\left.\begin{aligned} &\frac{d\mathbf{D}}{dt'} + \mathbf{D}\operatorname{div}\mathbf{u} - \mathbf{D}\nabla\,.\,\mathbf{u} + 4\pi\mathbf{C} = V\operatorname{curl}\mathbf{H}' \\ &\frac{d\mathbf{H}}{dt'} + \mathbf{H}\operatorname{div}\mathbf{u} - \mathbf{H}\nabla\,.\,\mathbf{u} + (\mu - 1)\left\{\frac{d\mathbf{H}'}{dt'} - \mathbf{H}'\nabla\,.\,\mathbf{u}\right\} = -V\operatorname{curl}\mathbf{E}' \end{aligned}\right\} \quad \text{...(97).}$$

It is interesting to notice that since

$$\frac{d\mathbf{G}'}{dt'} = (\mu - 1)\frac{d\mathbf{H}'}{dt'} + \mathbf{H}'\frac{d\mu}{dt'} = (\mu - 1)\left\{\frac{d\mathbf{H}'}{dt'} - \mathbf{H}'\operatorname{div}\mathbf{u}\right\},$$

the second of these equations can be put into the form

$$\frac{d\mathbf{H}}{dt'} + \mathbf{H}\operatorname{div}\mathbf{u} - \mathbf{H}\nabla\,.\,\mathbf{u} + \frac{d\mathbf{G}'}{dt'} - \mathbf{G}'\nabla\,.\,\mathbf{u} + \mathbf{G}'\operatorname{div}\mathbf{u} = -V\operatorname{curl}\mathbf{E}':$$

$$\therefore\ \frac{d\mathbf{G}}{dt'} + \mathbf{G}\operatorname{div}\mathbf{u} - \mathbf{G}\nabla\,.\,\mathbf{u} = -V\operatorname{curl}\mathbf{E}' \quad \text{......(98),}$$

and this form of the equation is unaltered by the change of hypothesis; cf. (26), § 14.

When $\mathbf{u}'' = 0$ we find

$$\left.\begin{aligned}\frac{d\mathbf{E}}{dt} + \mathbf{u}\operatorname{div}\mathbf{E} + (K-1)\left\{\frac{d\mathbf{E}'}{dt'} + \mathbf{E}'\operatorname{div}\mathbf{u} - \mathbf{E}'\nabla\,.\,\mathbf{u}\right\}&\\ + 4\pi\mathbf{C} = V\operatorname{curl}\mathbf{H}&\\ \frac{d\mathbf{H}}{dt} + \mathbf{u}\operatorname{div}\mathbf{H} + (\mu-1)\left\{\frac{d\mathbf{H}'}{dt'} - \mathbf{H}'\nabla\,.\,\mathbf{u}\right\} = -V\operatorname{curl}\mathbf{E}&\end{aligned}\right\}\ \ldots(99).$$

Hence, as before,

$$\left.\begin{aligned}\mathbf{E}'' &= \mathbf{E} + \frac{1}{V}[\mathbf{u}''\mathbf{H}]\\ \mathbf{H}'' &= \mathbf{H} + \frac{1}{V}[\mathbf{E}\mathbf{u}'']\end{aligned}\right\}\ \ldots\ldots\ldots\ldots\ldots(100),$$

and

$$\left.\begin{aligned}\mathbf{E}' &= \mathbf{E} + \frac{1}{V}[\mathbf{u}\mathbf{H}]\\ \mathbf{H}' &= \mathbf{H} + \frac{1}{V}[\mathbf{E}\mathbf{u}]\end{aligned}\right\}\ \ldots\ldots\ldots\ldots\ldots\ldots(101).$$

This theory gives the same explanation of Röntgen's experiment as that of §§ 25—27: when $\mu = 1$ the equations are independent of the hypothesis made as to the nature of magnetic media.

49. Let us consider the stresses in an isotropic but not necessarily homogeneous medium.

Taking, as the energy per unit volume,

$$W = \{\mathbf{E}^2 + (K-1)\mathbf{E}'^2 + \mathbf{H}^2 + (\mu-1)\mathbf{H}'^2\}/8\pi,$$

we find, with the former notation (§ 34),

$$\begin{aligned}\frac{1}{\omega}\frac{d}{dt'}(W\omega) = {}&\frac{1}{4\pi}\left\{\mathbf{E}\frac{d\mathbf{E}}{dt'} + (K-1)\mathbf{E}'\frac{d\mathbf{E}'}{dt'} + \mathbf{H}\frac{d\mathbf{H}}{dt'} + (\mu-1)\mathbf{H}'\frac{d\mathbf{H}'}{dt'}\right\}\\ &+ \frac{1}{8\pi}\frac{d\mu}{dt'}\mathbf{H}'^2\\ &+ \frac{1}{8\pi}\{\mathbf{E}^2 + (K-1)\mathbf{E}'^2 + \mathbf{H}^2 + (\mu-1)\mathbf{H}'^2\}\frac{1}{\omega}\frac{d\omega}{dt'}\\ = {}&\frac{1}{4\pi}\left\{(\mathbf{E}-\mathbf{E}')\frac{d\mathbf{E}}{dt'} + (\mathbf{H}-\mathbf{H}')\frac{d\mathbf{H}}{dt'}\right\} +\end{aligned}$$

$$+\frac{1}{4\pi}\mathbf{E}'\{-\mathbf{D}\operatorname{div}\mathbf{u}+\mathbf{D}\nabla.\mathbf{u}-4\pi\mathbf{C}+V\operatorname{curl}\mathbf{H}'\}$$

$$+\frac{1}{4\pi}\mathbf{H}'\{-\mathbf{H}\operatorname{div}\mathbf{u}+\mathbf{H}\nabla.\mathbf{u}+(\mu-1)\mathbf{H}'\nabla.\mathbf{u}-V\operatorname{curl}\mathbf{E}'\}$$

$$-\frac{\mu-1}{8\pi}\mathbf{H}'^2\operatorname{div}\mathbf{u}$$

$$+\frac{1}{8\pi}\{\mathbf{E}^2+(K-1)\mathbf{E}'^2+\mathbf{H}^2+(\mu-1)\mathbf{H}'^2\}\operatorname{div}\mathbf{u}$$

$$=\frac{1}{4\pi V}\left\{[\mathbf{Hu}]\frac{d\mathbf{E}}{dt'}+[\mathbf{uE}]\frac{d\mathbf{H}}{dt'}\right\}-\mathbf{E}'\mathbf{C}+\frac{V}{4\pi}\operatorname{div}[\mathbf{H}'\mathbf{E}']$$

$$+\frac{1}{8\pi}\{\mathbf{E}^2+(K-1)\mathbf{E}'^2-2\mathbf{DE}'+\mathbf{H}^2-2\mathbf{HH}'\}\operatorname{div}\mathbf{u}$$

$$+\frac{1}{4\pi}\{\mathbf{E}'.\mathbf{D}\nabla.\mathbf{u}+\mathbf{H}'.\mathbf{G}\nabla.\mathbf{u}\}\ \ldots\ldots\ldots\ldots\ldots\ldots(102).$$

This expression may be obtained from the corresponding expression (80) of § 34 by the addition of

$$\frac{1}{8\pi}\{2\mathbf{GH}'-(\mu-1)\mathbf{H}'^2-2\mathbf{HH}'\}\operatorname{div}\mathbf{u},$$

or
$$\frac{\mu-1}{8\pi}\mathbf{H}'^2\operatorname{div}\mathbf{u},\ \text{i.e.}\ \frac{\mu-1}{8\pi}\mathbf{H}'^2(u_x+v_y+w_z).$$

Thus we have an additional tension in the ether of $(\mu-1)\mathbf{H}'^2/8\pi$ parallel to each of the three axes, and an additional resultant force per unit volume whose component parallel to OX is

$$\frac{d}{dx}\left\{\frac{\mu-1}{8\pi}\mathbf{H}'^2\right\},$$

or
$$(\mu-1)\mathbf{H}'\mathbf{H}_x'/4\pi+\mu_x\mathbf{H}'^2/8\pi\ldots\ldots\ldots\ldots(103).$$

50. In addition to the Poynting flux $\frac{V}{4\pi}[\mathbf{E}'\mathbf{H}']$ and to the work $\mathbf{E}'\mathbf{C}$ done in maintaining conduction currents, we now have in all (cf. § 34) a force $-\frac{d\mathbf{P}}{dt'}$ occurring explicitly, and stresses in the ether represented by:

A tension parallel to OX of amount

$$\frac{1}{8\pi}\{\mathbf{E}^2 - 2YY' - 2ZZ' + (K-1)(X'^2 - Y'^2 - Z'^2)$$
$$+ \mathbf{H}^2 - 2MM' - 2NN' + 2(\mu - 1)L'^2\},$$

a shearing stress

$$\frac{1}{8\pi}\{Y'\mathfrak{Z} + \mathfrak{Y}Z' + M'\mathfrak{N} + \mathfrak{M}N'\},$$

and a couple

$$\frac{1}{4\pi}\{Y'Z - YZ' + M'N - MN'\} \quad \text{.........(104).}$$

Owing to the identity of equations (97), (98) with (21), (26), on which §§ 34, 35 depend, we may form the component Ξ {parallel to OX} of the resultant force per unit volume by adding the supplementary terms of (103) to the result (82) of § 35. We find

$$\rho X' + \tau L' + \frac{1}{V}(q\mathfrak{N} - r\mathfrak{M}) - K_x\mathbf{E}^2/8\pi + k\mathbf{H}'\mathbf{H}_x'$$
$$+ \frac{d}{dt'}(\mathfrak{P} - P) + \mathfrak{P}\,\text{div}\,\mathbf{u} + \mathbf{u}\mathbf{P}_x + \mathfrak{Q}\mathbf{u}_x \quad \text{...(105),}$$

where ρ, τ indicate densities in addition to those of the induced polarisations.

The resultant force on an uncharged stationary element of ether vanishes, as it should do.

51. When the ratio u/V is neglected we have

$$\Xi = \rho X + \tau L + (qN - rM)\mu/V - K_x\mathbf{E}^2/8\pi + k\mathbf{H}\mathbf{H}_x + (\mu K - 1)\frac{dP}{dt},$$

and

$$\mathbf{F} = \rho\mathbf{E} + \tau\mathbf{H} + \frac{1}{V}[\mathbf{C}\mathbf{G}] - \frac{1}{8\pi}\mathbf{E}^2 . \nabla K + \tfrac{1}{2}k\nabla\mathbf{H}^2 + (\mu K - 1)\frac{d\mathbf{P}}{dt}$$
$$\text{............(106),}$$

and the stresses (104) in the ether become:

Tensions represented by

$$\frac{1}{8\pi}\{K(X^2 - Y^2 - Z^2) + \mu(L^2 - M^2 - N^2) + (\mu - 1)(L^2 + M^2 + N^2)\},$$

shears of which an example is

$$\frac{1}{4\pi}\{KYZ+\mu MN\}:$$

and couples which vanish ....(107).

52. Let us now investigate the corresponding stresses in the material medium.

In an electrostatic field we may take μ as unity and the fundamental equations underlying the previous section become identical with those of §§ 34—37, when all magnetic effects are excluded. Thus the analysis of §§ 40—44, as applied to an electrostatic field, remains valid. As determined in § 41, the effect of the stresses in the ether is that there is a force $e\mathbf{E}$ upon an ion whose charge is e. The smallest volume of which account is taken in the theory of elasticity will contain an immense number of ions: and the consequence of the forces $e\mathbf{E}$ is the production of a resultant force, per unit volume, equal to

$$\mathbf{F}=\rho\mathbf{E}-\frac{1}{8\pi}\mathbf{E}^2.\nabla K \;..............(108),$$

and a force, per unit area of a bounding surface between two media, whose components are

$$\{K(X^2-Y^2-Z^2)/8\pi\}_1^2,\quad \sigma Y,\quad \sigma Z ....(109),$$

the axis of x being the normal drawn into the second medium.

When the dielectric is initially uncharged and is homogeneous, $\mathbf{F}$ vanishes and the force on the surface is

$$\{K(X^2-Y^2-Z^2)/8\pi\}_1^2,\quad 0,\quad 0,$$

i.e. is $\{K(2X_n^2-\mathbf{E}^2)/8\pi\}_1^2$ along the normal into the second medium, X_n standing for the electric force in the direction of that normal.

When the axes have general directions and the normal has direction cosines (l, m, n) the normal force is

$$\{2K(lX+mY+nZ)^2-K\mathbf{E}^2\}_1^2/8\pi \;......(110).$$

As already explained this normal force is equilibrated either by the elasticity of a solid medium, or, in an uncharged liquid, by additional hydrostatic pressure and by surface tension.

53. In a magnetostatic field the stresses in the ether are given by § 51 and consist of tensions,

$$\{\mu (L^2 - M^2 - N^2) + (\mu - 1)(L^2 + M^2 + N^2)\}/8\pi, \text{ etc.,}$$

and shears,

$$\{\mu MN\}/4\pi, \text{ etc.}$$

The couples vanish ....(111).

These give rise to an internal force per unit volume given by

$$\begin{aligned} \Xi &= \tau L + k\mathbf{H}\mathbf{H}_x \\ &= \tau L + k(LL_x + ML_y + NL_z) = \tau L + k\,.\,\mathbf{H}\nabla\,.\,L \quad \text{......(112).} \end{aligned}$$

On a magnetic molecule of moment $\mathbf{M}$ we should, according to Maxwell's theory, expect a force $\mathbf{M}\nabla\,.\,L$: cf. *Electricity and Magnetism*, §§ 389, 639. The formula for Ξ gives a force $\mathbf{G}'\nabla\,.\,L/4\pi$ per unit volume when the induced moment per unit volume is $\mathbf{G}'/4\pi$: thus the result is such as might be expected to follow from our hypothesis that the molecules behave as if independent of each other.

Further, at a surface of rapid though continuous transition to which OX is normal, Ξ is large and the normal force per unit area is, by the method of § 44,

$$\begin{aligned} \int dx\,\Xi &= \frac{1}{8\pi}\int dx\left(\frac{d}{dx}\{\mu(L^2 - M^2 - N^2) + (\mu - 1)(L^2 + M^2 + N^2)\}\right. \\ &\qquad\qquad \left. + \frac{d}{dy}\{2\mu LM\} + \frac{d}{dz}\{2\mu LN\}\right) \\ &= \{\mu(L^2 - M^2 - N^2) + (\mu - 1)(L^2 + M^2 + N^2)\}_1^2/8\pi \\ &= \{(2\mu - 1)L^2 - M^2 - N^2\}_1^2/8\pi \\ &= \{(2\mu - 1)L^2/8\pi\}_1^2; \end{aligned}$$

for $\qquad \{\mu L\}_1^2 = 4\pi v, \quad \{M\}_1^2 = 0, \quad \{N\}_1^2 = 0,$

where v is the surface density of permanent magnetism, if any is present.

Similarly

$$\begin{aligned} \int dx\,\mathrm{H} &= \{\mu LM/4\pi\}_1^2 \\ &= vM, \\ \int dx\,\mathrm{Z} &= vN. \end{aligned}$$

Hence when there is no distribution of permanent magnetism, we have

$$\mathbf{F} = \tfrac{1}{2}k\nabla\mathbf{H}^2 \equiv (k\mathbf{H}\mathbf{H}_x,\ k\mathbf{H}\mathbf{H}_y,\ k\mathbf{H}\mathbf{H}_z) \ldots\ldots (113),$$

and the force acting upon the matter at a surface of discontinuity is

$$\{(2\mu - 1)(lL + mM + nN)^2/8\pi\}_1^2,$$

in the direction (l, m, n) of the normal drawn into the second medium; the direction of the axes is here arbitrary.

Since

$$\{\mu(lL + mM + nN)\}_1^2 = 0,$$

$$\therefore\ \{\mu^2(lL + mM + nN)^2/8\pi\}_1^2 = 0.$$

Hence by subtraction the normal surface-force may be put into the form

$$-\{(\mu - 1)^2(lL + mM + nN)^2/8\pi\}_1^2,$$

or

$$-2\pi\{k^2(lL + mM + nN)^2\}_1^2 \ldots\ldots\ldots\ldots\ldots (114).$$

In the case of a liquid medium in equilibrium the force $\mathbf{F}\ (= \tfrac{1}{2}k\nabla\mathbf{H}^2)$ will involve at each point a hydrostatic pressure differing by a constant from $\tfrac{1}{2}k\mathbf{H}^2$. The normal surface-force will be counterbalanced either by effects due to surface-tension or by some additional hydrostatic pressure which is constant through the liquid.

Comparison of theoretical stresses with the results of observation.

54. We shall in most cases adopt as the standard for comparison the series of "Electrische Untersuchungen," published by Quincke in *Wiedemann's Annalen*: we shall consider in particular the experiments on liquid dielectrics of Vol. XIX. (pp. 705—782), with J. Hopkinson's correction applied in Vol. XXXII. (pp. 529—544), and the investigations concerning liquid magnetic media of Vol. XXIV. (pp. 347—416).

55. We shall begin with the experimental investigations of the stresses in an electrostatic field.

(α) Quincke measured by means of a balance the attraction between the two parallel plates of a condenser when separated by a liquid dielectric; and, on the hypothesis that the force was $K_p E^2/8\pi$ per unit area, determined the value of K_p. His apparatus is described in *Wied. Ann.* XIX. §§ 51, 52, pp. 707—717, and Fig. 22, Taf. VIII.

Quincke also determined the increase of pressure in an air-bubble in a liquid dielectric between the plates of the condenser used in the previous experiment; the plates were ·15 cm. apart, and the bubble extended from one to the other, occupying the central portion of the space between them; its diameter was between two and five cm., that of the plates being 8·5 cm. (*W. A.* XIX. § 53, pp. 718—726; Fig. 24, Taf. VIII.). On the assumption that this increase was equal to

$$(K_s - 1) E^2/8\pi$$

he obtained the value of K_s.

The comparison of K, K_p, K_s of p. 725 is not very satisfactory; K_s and K_p agree well enough, but K is decidedly smaller. When however the determinations of K have been corrected for the capacity of the wires (*W. A.* XXXII. pp. 531—534), the differences between K, K_p, K_s become extremely small: the ratios of K_p to K for five substances (p. 532) are 1·043, 1·002, 1·005, 0·985, 1·026.

It is easily seen that these surface-forces have the theoretical values given by the formula $\{K(X^2 - Y^2 - Z^2)/8\pi\}_1^2$ of §§ 43, 44, the axis of X being along the normal drawn into the second medium. In the first case there is no electric force within the condenser plate and there will be a tension of $KE^2/8\pi$ per unit area towards the other plate. In the second case the force perpendicular to the plates is tangential to the surface of the bubble and will have the same value E just within the bubble as just outside it. Thus the force at the surface in the direction of the normal drawn into the liquid is $-KE^2/8\pi + E^2/8\pi$: and the consequent increase of pressure will be $(K - 1) E^2/8\pi$. The liquid being uncharged, there will be no mechanical force (108) in its interior produced directly by the electric field.

(β) An interesting experiment was made by Quincke with the object of ascertaining the effect of the electric field upon the shape of an air-bubble in a liquid medium; he formed against the upper plate of the condenser above described an air-bubble whose rounded surface did not reach as far as the lower plate (*W. A.* XXIV. p. 376).

When the condenser was charged the bubble shewed conspicuous changes of form: it expanded in the direction of the lines of force and contracted in directions at right angles to them.

We shall not attempt an elaborate quantitative investigation, but shall consider the simple case in which the size of the bubble is small; its shape will then be approximately spherical, and, if the angle of contact is infinitesimal, its surface will just touch the upper plate. If the uniform field in the liquid, before the bubble makes its appearance, be of unit strength in the direction OX, it is easy to shew that the potentials inside and outside the bubble will become

$$\left.\begin{aligned} V_i &= -\frac{3K}{2K+1}\,x \\ V_0 &= -x - \frac{K-1}{2K+1}\,x\left(\frac{a}{r}\right)^3 \end{aligned}\right\}.$$

Thus the field inside the bubble is uniform; its strength will be denoted by E_i. Then at points where the lines of force cut the surface normally, the force just outside the bubble is E_i/K and, by (110), the surface-force inwards is

$$E_i^2/8\pi - K\left(\frac{E_i}{K}\right)^2 \Big/ 8\pi,$$

or

$$\frac{K-1}{8\pi K}\,E_i^2.$$

At points where the lines of force are tangential, the electric force inside and outside is continuous, and the surface-force inwards is

$$-E_i^2/8\pi + KE_i^2/8\pi,$$

or

$$\frac{K-1}{8\pi}\,E_i^2,$$

which is K times the force just obtained.

A similar result holds in the case of an ellipsoidal air-bubble, within which the field will also be uniform.

Hence the shape of the bubble of finite dimensions might be expected to change in the manner actually observed.

(γ) In the case of a solid dielectric, the results may be complicated by changes in the electric and elastic qualities of the substance. But in many cases, at any rate, the theoretical surface-forces suffice for a qualitative explanation. Examples may be found in Wiedemann's *Lehre von der Elektricität*, Band II. §§ 186—201: of these two may be taken as typical.

Quincke (*W. A.* X. pp. 161—190) took a glass vessel in the shape of a spherical bulb of about 5 cm. diameter at the end of a capillary tube, filled it with liquid, and observed the fall of this liquid in the capillary tube when the vessel was used as a condenser and charged.

The electric force within the glass being normal and denoted by E, that in the partially conducting liquid in contact with it will be extremely small and the surface-force will be $KE^2/8\pi$ in the direction of the normal drawn into the glass. This normal pressure on the glass at each surface will cause it to expand laterally: the diameter of the sphere and the volume of its interior will therefore increase, and the liquid in the capillary tube will fall. For further discussion see *W. A.* X. pp. 513—520: XI. pp. 771—780.

Righi (*Comptes Rendus*, 88, p. 1262) and Quincke (*W. A.* X. pp. 374—384) have shewn that a cylindrical glass tube increases in length when used as a condenser and charged. A somewhat similar experiment is due to Röntgen (*W. A.* XI. p. 786; 1880).

56. We now pass to the experimental determinations of stress in a magnetic field. In a uniform magnetic medium the field produces, according to our theory, a force per unit volume equal to $\frac{1}{2}k\nabla\mathbf{H}^2$ and at a surface between two such media a normal force $2\pi\{k^2(lL+mM+nN)^2\}_1^2$: cf. (113), (114), § 53. Hence at a boundary between the magnetic medium and air, the surface force bears to the hydrostatic pressure $\frac{1}{2}k\mathbf{H}^2$ caused by the internal force a ratio comparable in general with $2\pi k^2/\frac{1}{2}k$ or $(\mu-1)$.

Now in the case of all the liquid magnetic media investigated by Quincke this ratio is small. The greatest susceptibility is that of a solution of chloride of iron in methyl alcohol, and for this liquid we find (*W. A.* XXIV. Table 85, p. 385) that $(\mu - 1)/8\pi$ is given as equal to $414{\cdot}7 \times 10^{-10}$. It must however be remembered that Quincke's constants are calculated on the hypothesis that $(\mu - 1)\,\mathbf{H}^2/8\pi$ is the pressure in grammes weight per square centimetre (cf. *W. A.* XXIV. p. 383, (7) et seq.): if the pressure is reckoned in C. G. S. units, *as it should be*, the tabulated value of $(\mu - 1)/8\pi$ must be multiplied by 981. This fact is pointed out by Quincke.

Hence the greatest value of $(\mu - 1)$ is $1{\cdot}02 \times 10^{-3}$ or ·00102. The pressure due to surface-forces will therefore in no case be larger than about one thousandth part of that arising from the force in the interior, and will be inappreciable.

(δ) In § 65 (*W. A.* XXIV. pp. 362—366: Fig. 1, Taf. VII.) we have an account of experiments in which an air-bubble is blown in a liquid occupying the space between the horizontal surfaces of the poles of an electromagnet. These poles are 1·7 mm. apart and 140 mm. in diameter: the bubble is in contact with each of them, and has a diameter comparable with 25 mm.

Since the lines of force would crowd into the space between the poles, the magnetic force H_s at the distant external surface of the liquid would be very small compared with its value H at the air-bubble. According to the present theory, as given in § 53, the liquid will be impelled from places of less magnetic force to places of greater intensity and the additional hydrostatic pressure in the liquid at the bubble should be

$$\frac{\mu - 1}{8\pi}(H^2 - H_s^2) \quad \text{or} \quad \frac{\mu - 1}{8\pi} H^2.$$

The additional pressure within the air-bubble (for which $\mu = 1$) should thus, in the absence of appreciable surface-forces, be $(\mu - 1)\,H^2/8\pi$.

A pressure is observed which is in agreement with this formula, and may be used for the determination of μ.

(ϵ) Quincke also uses (*W. A.* XXIV. pp. 369—374: Figs. 7, 8,

Taf. VIII.) a U-shaped tube, containing the magnetic liquid as a manometer. One vertical arm passes between the poles of an electromagnet arranged in such a manner that the lines of force at the surface of the liquid shall be horizontal. This surface rises when the current is turned on, and by using the same electromagnet as in the previous experiment, it may be shewn that the increase of pressure due to the field is the same as it was in the case of the air-bubble. This is in accordance with the theory of § 53, and the results are given in *W. A.* XXIV. Table 83, p. 373.

(ζ) In § 68 (pp. 374, 5) the poles of the electromagnet have their faces horizontal, the lines of force between them being vertical; one vertical arm of the U-tube of the previous experiment is passed up through a vertical hole in the axis of the lower pole in such a manner that the surface of the liquid in the tube lies in the magnetic field and is cut at right angles by the lines of force.

The surface of the liquid now rises when the current is turned on: as we should expect, the indicated increase of pressure is equal to that observed when the lines of force were parallel to the surface and the conditions were otherwise the same (Table 84).

(η) Quincke made an interesting experiment (*l.c.*, p. 376) upon the effect of the field on an air-bubble lying in a magnetic liquid between the two horizontal surfaces of the poles of an electromagnet. The bubble touched only the upper pole, its rounded surface approaching near to the lower pole. When the field was excited, the shape of the bubble was not affected: the increase of pressure at the surface of the bubble was the same whether the lines of force were parallel to the tangent-plane or perpendicular to it.

(θ) Other investigations were made upon the shape of drops hanging from a rod (pp. 376, 7). Even in a field of 12,000 C.G.S. units, a drop of a concentrated magnetic liquid, when examined by a kathetometer microscope, did not change its shape.

Under the same conditions a drop of iron amalgam preserved

the same diameters parallel and perpendicular to the lines of force; but changed its shape at once if the field were not of uniform strength. For this amalgam in Quincke's Table 85 $(\mu-1)/8\pi$ is given as 160·3: hence in C.G.S. units $(\mu-1)=\cdot 0004$, and the surface-forces are inappreciable.

(ι) The term $\frac{1}{2}k\nabla\mathbf{H}^2$ in $\mathbf{F}$ is illustrated by the fact that drops of solution of chloride of iron in alcohol move in diamagnetic olive-oil of the same density towards places of stronger magnetic force (Matteucci, *Comptes Rendus*, 36, p. 917, 1853).

(κ) Consider the force exerted between two equal longitudinally magnetised cylindrical electromagnets of great length with plane ends at right angles to their axes. The electromagnets are placed with their axes in the same straight line and with opposite poles facing one another, their distance apart being small compared with the diameter of either.

The field in the air-gap may be treated as uniform; and, if H_2, H_1 denote the magnetic force in the air-gap and inside the iron, we shall have $H_2 = \mu H_1$. The force per unit area of the surface is $2\pi k^2 H_1^2$ in the direction of the normal drawn into the air. The force per unit volume of the interior is $\frac{1}{2}k\nabla\mathbf{H}^2$; and if $\mathbf{H}$ is negligible at the distant end of the magnet, the resultant effect is to give an integral force of $\frac{1}{2}kH_1^2$ per unit area of cross section in the same direction as the above force $2\pi k^2 H_1^2$. The sum of these forces is $\frac{1}{2}k(1+4\pi k)H_1^2$, or $(\mu^2-\mu)H_1^2/8\pi$. This result is equivalent to $\{\mu H^2/8\pi\}_1^2$ which corresponds, as it should, to the electric surface-force $\{KE^2/8\pi\}_1^2$. For soft iron $1/\mu$ is small and the experimental comparison by E. Taylor Jones of the surface force with $\mu^2 H_1^2/8\pi$ (*Phil. Mag.* 39, p. 254, 1895) proves our expression correct to one per cent.

(λ) In the case of most of these magnetic experiments it will be seen that an explanation might be effected if we were to assume that the forces due to a magnetic field were similar in character to those of §§ 42, 43 due to an electric field. We should then have a force at a bounding surface equal to $\{\mu(L^2-M^2-N^2)/8\pi\}_1^2$ in the direction of OX, the normal drawn into the second medium, and zero force in the interior.

In (θ) if the drop of magnetic liquid in a uniform field be treated as spheroidal in shape, the force inside the drop will be uniform: let us denote it by H_1.

The external magnetic force at points where the normal is parallel to H_1 will be μH_1, and there would be an outward surface-thrust of $(\mu^2 - \mu) H_1^2/8\pi$. Where the tangent planes are parallel to H_1, the external magnetic force will be H_1 and the outward surface-thrust would be $(\mu - 1) H_1^2/8\pi$. The ratio of these stresses would be $\mu : 1$, and their difference would be inappreciable.

In the case of (δ), (ϵ), (ζ), (η) explanation on these lines is simple.

A crucial experiment may be made by observing the change of volume due to the pressure produced by the field in a magnetic liquid in an open receptacle. We shall consider in particular a thermometer-shaped vessel placed symmetrically with its capillary tube vertical and its bulb in the centre of an intense field of horizontal force between the poles of a powerful electromagnet.

At the free surface of the liquid in the capillary tube the magnetic force H_s will be horizontal in direction, and the expression

$$2\pi \{k^2 (lL + mM + nN)\}^{\frac{1}{2}}$$

for the normal surface-stress will be small. When the faces of the poles are near together and the field throughout the bulb may be treated as of uniform intensity H, then the additional hydrostatic pressure produced by the force $(\mu - 1) \nabla \mathbf{H}^2/8\pi$ per unit volume will be $(\mu - 1)(H^2 - H_s^2)/8\pi$. Under this pressure the liquid will contract, and the walls of the bulb will yield slightly: thus the column in the capillary tube will fall.

If, however, the stresses were analogous with those of an electric field they would give merely a force $(\mu - 1) H_s^2/8\pi$ per unit area acting vertically upwards at the surface of the liquid in the tube. Hence this column would rise.

If the magnetic liquid is surrounded by a column of water and no diffusion occurs, then neglecting the infinitesimal susceptibility of the water, the result will be unaffected by its presence. When the magnetic liquid is such that partial diffusion occurs, and the permeability increases continuously as we pass from the top of the column of water down to the magnetic liquid, then our theory still gives zero force at the free surface in the capillary tube and $(\mu - 1)\nabla\mathbf{H}^2/8\pi$ at each point in the interior. Whatever the law of diffusion this will act vertically downwards in the capillary tube and the column will fall.

On the other hand, the stresses of the electric type would consist of a very small force $(\mu_s - 1) H_s^2/8\pi$ upwards at the upper surface of the water column, μ_s being the permeability of the water there with its trace of magnetic liquid: and in the interior a force $-\mathbf{H}^2\nabla\mu/8\pi$ by (84). Since μ increases with the depth below the free surface, this force would also act vertically upwards within the capillary tube. Thus the column would rise.

We may apply a further test to our theory by ascertaining the changes of volume corresponding to different positions of the vessel in the field. If the bulb is lowered until it is below the more intense field between the poles, while the magnetic liquid at the bottom of the capillary tube is still in that field, then the force $(\mu - 1)\nabla\mathbf{H}^2/8\pi$ will act upwards on that part of the liquid which is below the centre of the field: and a diminution of the internal pressure will result. Thus we should expect the column in the capillary tube to rise.

I lately ventured* to ask Prof. Quincke for some details in connection with his observations in relation to change of volume (*W. A.* XXIV. p. 380, § 68), and it is through very great kindness on his part, for which I cannot be too grateful, that I am enabled to state the results of some additional experiments recently made† by him with the object of testing the accordance of the above theory with fact.

For one set of investigations he took a thermometer-shaped vessel whose bulb was a cylinder 4 mm. in length and 35 mm. in the diameter of its circular cross-section: the diameter of the capillary tube was ·334 mm. and its height 230 mm. The ends of the poles of the electromagnet were squares of

* In December 1899.

† In February and March 1900.

51 mm. × 51 mm., and their distance apart was either 5·2 mm. or 6·745 mm. The magnetic liquid used was a strong aqueous solution of ferric chloride whose specific gravity σ was 1·4933.

Quincke performed the preliminary experiment of filling the bulb and part of the tube with water, introducing a column of mercury above the column of water, and noting the change in the position of its extremity produced by inverting the vessel: this gave the ratio of the increase of volume to the decrease of pressure.

Observations of the type (ϵ) above, with a U-tube used as a manometer, were made with the ferric chloride solution. The height of h cm. through which the column of liquid rose was connected with H, the strength of the field where uniform, and with Δp_m, the corresponding increase of pressure in grammes weight per square centimetre, by the relations

$$\Delta p_m = h\sigma = 320{\cdot}9 \times 10^{-10} H^2.$$

Let the distance between the poles of the electromagnet in millimetres be denoted by a, let z mm. be the rise in height in the capillary tube when the thermometer contains the solution of ferric chloride and is placed in the field, and let Δp be the corresponding increase of pressure in grammes weight per square centimetre. The preliminary experiment shewed that

$$\Delta p = -5{\cdot}347 z.$$

When the bulb was in the centre of the field the value of the force H_s at the top of the column was very small compared

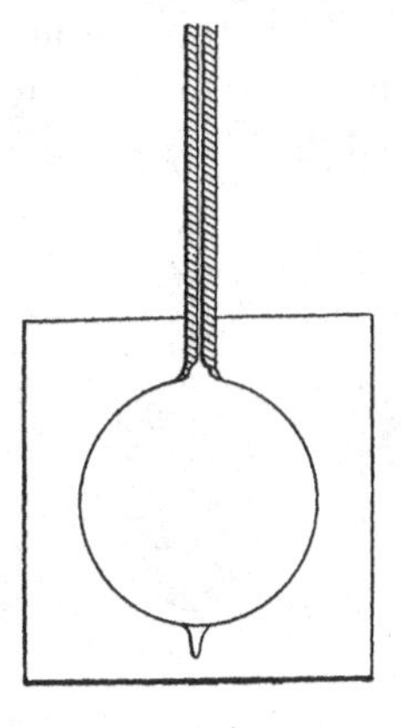

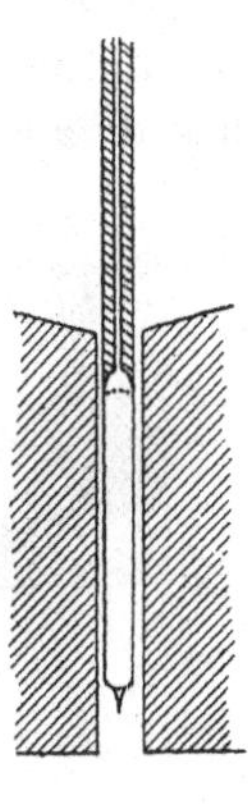

with H, so that the theoretical increase in pressure became approximately $(\mu - 1) H^2/8\pi$, the pressure in the manometer tube.

The observations gave :—

a	H	h	z	Δp_m	Δp
mm.	C. G. S.	cm.	mm.	$= h\sigma$	$= -5{\cdot}347z$
5·2	17,880	6·87	− 1·86	10·26	9·94
5·2	11,090	2·64	− 0·84	3·94	4·49
6·745	17,390	6·50	− 1·78	9·705	9·517
6·745	9,768	2·042	− 0·43	3·048	2·300

It will be seen from the sign of z that the column falls, as it should do. Moreover, the agreement in value between Δp_m and Δp is as close as the experimental difficulties would permit us to expect.

Investigations of a similar character with a larger vessel gave the results:

Δp_m	Δp
3·27	2·99
9·85	10·69

Quincke made experiments in which the thermometer occupied the five different positions indicated in the diagram : the observed values of z were − 1·8, − 1·2, − 0·8, + 0·3, + 1·1. In

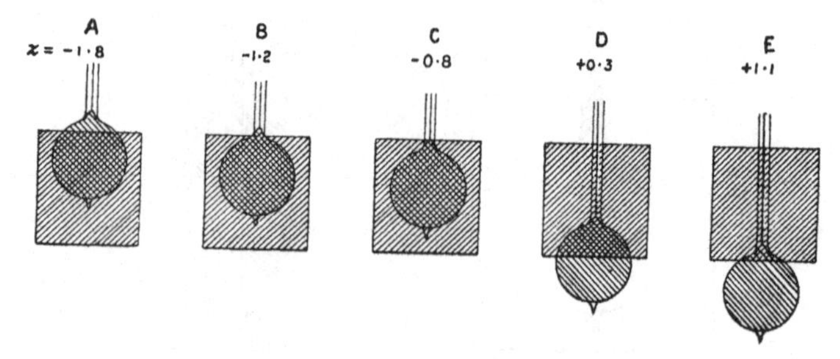

these experiments the bulb was filled with the solution of ferric chloride and the capillary tube contained water, the boundary being just above the bottom of the tube. It will be noticed that, owing to the closeness of the poles, the strength of the field must diminish rather suddenly at the margin of the region actually between them.

The numerical value of z accordingly (see A, B, C in the figure) becomes rapidly smaller as the top of the bulb passes below the level of the top of the poles and the field in which the magnetic liquid lies becomes more nearly uniform.

On the other hand the changes in the value of z as the bulb emerges below the poles will be less rapid (cf. D, E). For z indicates the average pressure in the bulb: and that part of the solution which is in the region of uniform force H will be subject merely to atmospheric pressure, while the pressure in the portion which is below the uniform region will be less than this by an amount increasing to nearly

$$(\mu - 1)\, H^2/8\pi,$$

at a point where the magnetic force is small.

Thus in the case of liquid media the agreement of experiment with theory seems to be complete.

PART IV.

PROPAGATION OF PLANE WAVES IN MOVING MEDIA WHEN THE POLARISATION IS CONTINUOUS, NOT MOLECULAR.

57. WE shall for the present suppose that the velocity $\mathbf{u}$ of any dielectric is independent both of the time and the coordinates: accordingly no allowances for rotation of the axes will be necessary. The dielectric will in all cases be uncharged.

The first case that will be considered is that in which, the ether being at rest, the origin is taken at a point fixed in the ether, and the axis OX is perpendicular to the wave-front. With the usual notation, we shall have, to axes fixed in the ether, if $\mu = 1$,

$$\left.\begin{aligned} \frac{d\mathbf{D}}{dt} &= \ \ V \operatorname{curl} \mathbf{H} \\ \frac{d\mathbf{H}}{dt} &= -V \operatorname{curl} \mathbf{E} \end{aligned}\right\} \qquad (115)$$

with

$$\left.\begin{aligned} \mathbf{D} &= \mathbf{E} + (K-1)\mathbf{E}' \\ \mathbf{E}' &= \mathbf{E} + \frac{1}{V}[\mathbf{uH}] \end{aligned}\right\} \qquad (116),$$

$$\operatorname{div} \mathbf{D} = 0, \quad \operatorname{div} \mathbf{H} = 0 \qquad (117).$$

If then the velocity of propagation be U, and we take

$$\frac{X}{\xi} = \frac{Y}{\eta} = \frac{Z}{\zeta} = \frac{L}{\lambda} = \frac{M}{\mu} = \frac{N}{\nu} = e^{is(Ut-x)},$$

the equations of the field are

$$\left.\begin{aligned} U\{K\xi+(K-1)(v\nu-w\mu)/V\}&=0,\\ U\{K\eta+(K-1)(w\lambda-u\nu)/V\}&=V\nu,\\ U\{K\zeta+(K-1)(u\mu-v\lambda)/V\}&=-V\mu \end{aligned}\right\};$$

$$\left.\begin{aligned} U\lambda&=0,\\ U\mu&=-V\zeta,\\ U\nu&=V\eta \end{aligned}\right\}.$$

Hence **D**, **H** lie in the wave-front, at right angles to one another, and

$$KU^2-(K-1)\,uU=V^2=KV'^2 \qquad (118),$$

where V' is the velocity of propagation in the medium at rest: therefore neglecting squares

$$U=V'+u(K-1)/2K \qquad (119).$$

Putting $x=x'+ut$, where x', y', z' are referred to axes fixed in the matter, the exponential becomes

$$e^{is\,(Ut-ut-x')},$$

and the velocity of propagation relative to the matter is

$$U-u \text{ or } V'-u(K+1)/2K \qquad (120).$$

The correct value of U as given by Fresnel's coefficient should be, cf. (119),

$$V'+u(K-1)/K.$$

58. The theory of Fresnel that in any body whose refractive index is m and velocity in space **u**, the ether is carried with velocity $(m^2-1)\,\mathbf{u}/m^2$, suggests the introduction of a similar hypothesis in the present case.

We shall accordingly suppose that when a medium has velocity **u** relative to the free ether at a great distance, the ether contained in the body has velocity $\alpha\mathbf{u}$, when α is some coefficient dependent on the medium in question.

Let $\alpha=1-\beta$, so that the velocity of the medium when referred to the contained ether is $\beta\mathbf{u}$. Now in the previous analysis in § 11 of the motion of bodies through ether, it is the ether which has been taken as "at rest," i.e. as affording the origin to which the motion is referred.

From this point of view, the formulae for continuous non-molecular polarisation,

$$\mathbf{E}' = \mathbf{E} + \frac{1}{V}[\mathbf{uG}], \qquad \mathbf{H}' = \mathbf{H} - \frac{1}{V}[\mathbf{uD}],$$

where $$\mathbf{D} = \mathbf{E} + (K-1)\,\mathbf{E}', \qquad \mathbf{G} = \mathbf{H} + (\mu - 1)\,\mathbf{H}',$$

have been developed: in these equations $\mathbf{u}$ is the velocity of the material medium relative to the ether. The analysis of §§ 11, 13 is accordingly applicable under the present conditions, provided that, when the velocity of the medium is $\mathbf{u}$ when referred to the ether at a great distance, we take

$$\mathbf{E}' = \mathbf{E} + \frac{1}{V}[\beta\mathbf{uG}], \qquad \mathbf{H}' = \mathbf{H} - \frac{1}{V}[\beta\mathbf{uD}] \quad \text{.........(121)},$$

where $\mathbf{E}$, $\mathbf{H}$ denote the forces at a point fixed in the ether, and $\mathbf{E}'$, $\mathbf{H}'$ at a point in the material medium.

The equations of the field, complete as far as aberration is concerned, when referred to axes fixed in the material medium, thus become

$$\left.\begin{aligned} \frac{d\mathbf{D}}{dt'} &= V \operatorname{curl} \mathbf{H}', \\ \frac{d\mathbf{H}}{dt'} &= -V \operatorname{curl} \mathbf{E}' \end{aligned}\right\};$$

$$\mathbf{D} = \mathbf{E} + (K-1)\,\mathbf{E}', \quad \operatorname{div} \mathbf{D} = 0, \quad \operatorname{div} \mathbf{H} = 0,$$

$$\mathbf{E}' = \mathbf{E} + \frac{\beta}{V}[\mathbf{uH}], \quad \mathbf{H}' = \mathbf{H} - \frac{\beta}{V}[\mathbf{uD}].$$

At a bounding surface between two media we shall suppose that the velocities $\mathbf{u}_1$, $\mathbf{u}_2$ of the two media are equal, and that K and β may change discontinuously. The velocities of the ether will then be discontinuous, but in our present state of ignorance, there is no insuperable difficulty in supposing the ether to have properties not possessed by ordinary matter.

By regarding the interface as the limiting case of a thin

layer of continuous transition and taking Oz perpendicular, we obtain in the same manner as on former occasions,

$$\begin{aligned} \{X'\}_1^2 &= 0, \quad \{Y'\}_1^2 = 0, \\ \{L'\}_1^2 &= 0, \quad \{M'\}_1^2 = 0, \\ \{Z + (K-1) Z'\}_1^2 &= 0, \quad \{N\}_1^2 = 0, \end{aligned}$$

the system containing no charges.

The velocity of propagation of a plane wave whose direction-cosines are l, m, n when referred to axes fixed in the material medium will be given by (120): it is

$$V' - \beta (ul + vm + wn)(K+1)/2K.$$

But since the velocity of light relative to the free ether at a distance is, by Fresnel's law, to be

$$V' + (ul + vm + wn)(K-1)/K,$$

or relative to the matter

$$V' - (ul + vm + wn)/K \quad \text{..............(122)};$$

$$\therefore \ \beta (K+1)/2K = 1/K.$$

$$\therefore \ \beta = \frac{2}{K+1}, \text{ and } \alpha = \frac{K-1}{K+1} \quad \text{......... (123)}.$$

59. Let us now consider the propagation relative to a uniform material medium of a disturbance emanating from a point. Denoting the velocity (122) of propagation by U' and $1/K$ by γ, the region of disturbance at the time t will be enveloped by the plane whose equation is

$$lx' + my' + nz' = U't = \{V' - \gamma (ul + vm + wn)\} t,$$

where l, m, n are subject to the relation

$$l^2 + m^2 + n^2 = 1.$$

Hence we introduce the quantity λ satisfying the equations

$$\left.\begin{aligned} x' + \gamma ut &= \lambda l \\ y' + \gamma vt &= \lambda m \\ z' + \gamma wt &= \lambda n \end{aligned}\right\}$$

and obtain $\lambda = V't$.

$$\therefore \ \frac{x'}{lV' - \gamma u} = \frac{y'}{mV' - \gamma v} = \frac{z'}{nV' - \gamma w} = t \text{.....(124)}.$$

Now the ray corresponding to the plane

$$lx' + my' + nz' = U't$$

meets it in that point on the plane which is first reached by the disturbance; it is therefore the point of contact of the plane with the envelope just investigated.

Hence the components of the velocity of ray-propagation relative to the material medium are

$$lV' - u/K, \quad mV' - v/K, \quad nV' - w/K \;\ldots\ldots (125).$$

Consider now the direction of the rays in the case of our former hypothesis of a stationary ether and a molecularly polarised material medium moving with velocity $\mathbf{u}$. The results (62), (63) of § 22 prove that when squares of u/V are neglected and when $\mathbf{u}$ is not necessarily along OX, the velocity of propagation of plane waves whose direction-cosines are (l, m, n), will be

$$U = V' + (lu + mv + nw)(K - 1)/K;$$

this velocity is referred to the free ether at a great distance.

The forces in the wave will be proportional to

$$e^{2\pi i \{Ut - lx - my - nz\}/\lambda};$$

if then we refer to axes moving with the material medium and replace x, y, z by $x' + ut$, $y' + vt$, $z' + wt$, we obtain the factor

$$e^{2\pi i \{(U - lu - mv - nw)t - lx' - my' - nz'\}/\lambda}.$$

Thus the velocity of wave propagation relative to the material medium is

$$U - (lu + mv + nw) \text{ or } V' - (lu + mv + nw)/K:$$

this is the same velocity as U' above. Hence also the direction and velocity of the rays given by the hypotheses of § 14 and of § 58 are the same.

Reflection and refraction at a plane bounding surface between two media possessing the same uniform velocity.

60. Let a point in the interface be taken as origin and the normal drawn into the first medium as the axis OZ, the origin and axes moving with the uniform velocity $\mathbf{u}$ of the system.

Let us suppose that in the first medium plane waves are being propagated in which the forces $\mathbf{E}'$, $\mathbf{H}'$, acting at points fixed relative to the axes, are proportional to $e^{is(Ut+lx+nz)}$.

For the waves reflected into the first medium $\mathbf{E}'$, $\mathbf{H}'$ may be assumed to vary as $e^{is_1(U_1t+l_1x+m_1y-n_1z)}$; and in the refracted wave we shall take as the corresponding exponential factor

$$e^{is_2(U_2t+l_2x+m_2y+n_2z)}.$$

At $z=0$ the indices of the exponentials must agree: hence

$$\left.\begin{aligned} sU &= s_1U_1 = s_2U_2 \\ sl &= s_1l_1 \;\; = s_2l_2 \\ 0 &= s_1m_1 = s_2m_2 \end{aligned}\right\} \quad \ldots\ldots\ldots\ldots\ldots\ldots(126).$$

Let V_1, V_2 denote the velocities of wave-propagation of light in the media when stationary; then, since m_1 and m_2 vanish, we deduce from (122),

$$\left.\begin{aligned} U &= V_1 + \gamma_1(lu+nw) \\ U_1 &= V_1 + \gamma_1(l_1u-n_1w) \\ U_2 &= V_2 + \gamma_2(l_2u+n_2w) \end{aligned}\right\};$$

$\therefore$ by (126),

$$\begin{aligned} s\{V_1+\gamma_1(lu+nw)\} &= s_1\{V_1+\gamma_1(l_1u-n_1w)\} \\ &= s_2\{V_2+\gamma_2(l_2u+n_2w)\}. \end{aligned}$$

The component velocities of the rays in the media are

$$\begin{aligned} &-V_1l-u/K_1, \quad -v/K_1, \quad -V_1n-w/K_1, \\ &-V_1l_1-u/K_1, \quad -v/K_1, \quad +V_1n_1-w/K_1, \\ &-V_2l_2-u/K_2, \quad -v/K_2, \quad -V_2n_2-w/K_2. \end{aligned}$$

Hence we have sufficient equations to determine the directions of the rays. If we required the intensities of the forces we should have to make use of the remaining surface conditions.

Now if the problem were investigated on the former hypothesis of molecular polarisation the equations giving the directions of the rays would be identical with those just found; but the surface conditions are somewhat different owing to the dissimilarity of the formulae

$$\mathbf{H}'=\mathbf{H}-\frac{1}{V}[\mathbf{uE}], \quad \mathbf{H}'=\mathbf{H}-\frac{\beta}{V}[\mathbf{uD}]$$

in the two cases. Hence the directions of the rays according to the two hypotheses are the same, but the intensities and planes of polarisation differ.

Now, as we have seen (p. 27), Lorentz has shewn that, if squares be neglected, the ordinary laws of reflection and refraction hold for rays of light in drifting media in which the polarisation is molecular: hence a similar theorem is true in the case of continuous polarisation if the hypothesis of § 58 be made.

It is also clear that this latter hypothesis will account for the ordinary facts of stellar aberration. For at the highly attenuated outer limits of the atmosphere, $\alpha \equiv (K-1)/(K+1)$ will be very small, and the velocity $\alpha\mathbf{u}$ with which the ether is carried along by the earth's motion will also be small. Thus on either hypothesis the ether at a very great distance from the earth is at rest: and our argument shews that such results as those of § 20 are valid under the present conditions.

Röntgen's experiment.

61. With the same axes as in § 26, the velocity $\mathbf{u}''$ of the centre of the sphere being (A, B, C) referred to the "free ether," the velocity $\mathbf{u}$ of any point (x, y, z) of the sphere will be $(A - ny, B + nx, C)$.

The equations of § 12, referred to axes moving with the uniform velocity $\mathbf{u}''$, will become, since $\rho = 0$, $\tau = 0$, $\mu = 1$ in the present case,

$$\left.\begin{aligned} \frac{d\mathbf{D}}{dt} + \mathbf{u}''\nabla \, . \, \mathbf{D} \equiv \frac{d\mathbf{D}}{dt''} &= \ \ V \operatorname{curl} \mathbf{H}'' \\ \frac{d\mathbf{H}}{dt} + \mathbf{u}''\nabla \, . \, \mathbf{H} \equiv \frac{d\mathbf{H}}{dt''} &= - V \operatorname{curl} \mathbf{E}'' \end{aligned}\right\}.$$

Since the conditions are steady, $\dfrac{d}{dt''} = 0$ and curl $\mathbf{E}''$, curl $\mathbf{H}''$ vanish: thus $\mathbf{E}''$, $\mathbf{H}''$ are each derivable from a potential.

Let suffixes 1, 2 distinguish quantities inside and outside the sphere. Then if $\mathbf{u}$, $\mathbf{u}''$ were zero, we should have

$$\mathbf{E}_1 = \nabla \frac{3}{K+2} z, \qquad \mathbf{E}_2 = \nabla \left\{ z - \frac{K-1}{K+2} \left(\frac{a}{r} \right)^3 z \right\},$$

and $\mathbf{H}_1$, $\mathbf{H}_2$ would vanish. Hence when $\mathbf{u}/V$, $\mathbf{u}''/V$ are small finite quantities, we shall have

$$\mathbf{E}_1'' = \nabla \left\{ \frac{3}{K+2} z + \phi_1 \right\}$$

$$\mathbf{E}_2'' = \nabla \left\{ z - \frac{K-1}{K+2} \left(\frac{a}{r} \right)^3 z + \phi_2 \right\}$$

$$\mathbf{H}_1'' = \nabla \omega_1$$

$$\mathbf{H}_2'' = \nabla \omega_2,$$

where ϕ_1, ϕ_2, ω_1, ω_2 are small functions.

Now the velocity of the centre of the sphere (our origin), relative to the ether at the point (x, y, z) within the sphere, is

$$\mathbf{u}'' - \alpha \mathbf{u},$$

or

$$(\beta A + \alpha n y,\ \beta B - \alpha n x,\ \beta C):$$

hence

$$\mathbf{E}_1'' = \mathbf{E}_1 + \frac{1}{V} [(\mathbf{u}'' - \alpha \mathbf{u}) \mathbf{H}],$$

$$\mathbf{H}_1'' = \mathbf{H}_1 - \frac{1}{V} [(\mathbf{u}'' - \alpha \mathbf{u}) \mathbf{D}].$$

Also the velocity $\mathbf{u}$ exceeds the velocity $\alpha \mathbf{u}$ of the ether by $\beta \mathbf{u}$: so that

$$\mathbf{E}' = \mathbf{E} + \frac{\beta}{V} [\mathbf{uH}],$$

$$\mathbf{H}' = \mathbf{H} - \frac{\beta}{V} [\mathbf{uD}].$$

As before (p. 36), if we neglect squares,

$$\mathbf{E}'' = \mathbf{E} = \mathbf{E}',$$

$$\therefore\ \mathbf{D} = K \mathbf{E}'';$$

and since $\operatorname{div} \mathbf{D} = 0$,

$$\therefore\ \nabla^2 \phi_1 = 0, \quad \nabla^2 \phi_2 = 0.$$

The surface conditions are that the tangential components of $\mathbf{E}'$, $\mathbf{H}'$ shall be continuous, as well as the normal components of $\mathbf{D}$, $\mathbf{H}$.

Hence $K \dfrac{d\phi_1}{dr} = \dfrac{d\phi_2}{dr}$, and the tangential components of $\nabla \phi_1$,

$\nabla\phi_2$ are continuous. Thus ϕ_1, ϕ_2 are electrostatic potentials due to a dielectric sphere without charge,

$$\therefore \quad \phi_1 = 0, \quad \phi_2 = 0.$$

Then, since

$$\mathbf{H}_1 = \mathbf{H}_1'' + \frac{K}{V}[(\mathbf{u}'' - \alpha\mathbf{u})\,\mathbf{E}_1],$$

$$\therefore \ \operatorname{div}\mathbf{H}_1 = \operatorname{div}\mathbf{H}_1'' + \frac{K}{V}\{\mathbf{E}_1 \operatorname{curl}(\mathbf{u}'' - \alpha\mathbf{u}) - (\mathbf{u}'' - \alpha\mathbf{u}) \operatorname{curl}\mathbf{E}_1\}.$$

Now $\quad \operatorname{curl}(\mathbf{u}'' - \alpha\mathbf{u}) = (0, 0, -2\alpha n),$

$$\operatorname{curl}\mathbf{E}_1 = 0, \ \operatorname{div}\mathbf{H}_1 = 0, \ Z_1 = 3/(K+2);$$

$$\therefore \quad 0 = \nabla^2\omega_1 - \frac{6K}{K+2}\frac{\alpha n}{V} \ldots\ldots\ldots\ldots\ldots\ldots(127).$$

Similarly $\quad \mathbf{H}_2 = \mathbf{H}_2'' + \frac{1}{V}[\mathbf{u}''\mathbf{E}_2],$

and $\quad \operatorname{div}\mathbf{H}_2 = 0, \ \operatorname{curl}\mathbf{u}'' = 0, \ \operatorname{curl}\mathbf{E}_2 = 0,$

$$\therefore \quad 0 = \nabla^2\omega_2 \ldots\ldots\ldots\ldots\ldots\ldots(128).$$

Further, remembering that outside the sphere $\alpha = 0$, $\beta = 1$, we have both inside and outside,

$$\mathbf{H}' = \mathbf{H} - \frac{1}{V}[\beta\mathbf{u}\mathbf{D}],$$

$$\mathbf{H}'' = \mathbf{H} - \frac{1}{V}[(\mathbf{u}'' - \alpha\mathbf{u})\,\mathbf{D}];$$

$$\therefore \quad \mathbf{H}' = \mathbf{H}'' - \frac{1}{V}[(\mathbf{u} - \mathbf{u}'')\,\mathbf{D}];$$

$$\therefore \quad L' = \frac{d\omega}{dx} - nx\mathfrak{Z}/V, \quad M' = \frac{d\omega}{dy} - ny\mathfrak{Z}/V,$$

$$N' = \frac{d\omega}{dz} + (ny\mathfrak{Y} + nx\mathfrak{X})/V.$$

The continuity of the tangential $\mathbf{H}'$ gives

$$\frac{L_2' - L_1'}{x} = \frac{M_2' - M_1'}{y} = \frac{N_2' - N_1'}{z};$$

$$\therefore \frac{1}{x}\left\{\frac{d(\omega_2-\omega_1)}{dx}-\frac{nx}{V}(Z_2-KZ_1)\right\}$$

$$=\frac{1}{y}\left\{\frac{d(\omega_2-\omega_1)}{dy}-\frac{ny}{V}(Z_2-KZ_1)\right\}$$

$$=\frac{1}{z}\left\{\frac{d(\omega_2-\omega_1)}{dz}+\frac{nx}{V}(X_2-KX_1)+\frac{ny}{V}(Y_2-KY_1)\right\},$$

or
$$\frac{1}{x}\frac{d(\omega_2-\omega_1)}{dx}=\frac{1}{y}\frac{d(\omega_2-\omega_1)}{dy}$$

$$=\frac{1}{z}\frac{d(\omega_2-\omega_1)}{dz}+\frac{n}{Vz}\{(xX_2+yY_2+zZ_2)$$

$$-K(xX_1+yY_1+zZ_1)\}.$$

Now the normal electric polarisation is continuous, so that the quantity multiplied by n/Vz may be neglected.

Hence
$$\frac{d(\omega_2-\omega_1)}{xdx}=\frac{d(\omega_2-\omega_1)}{ydy}=\frac{d(\omega_2-\omega_1)}{zdz};$$

$$\therefore \frac{d}{d\theta}(\omega_2-\omega_1)=0,$$

and
$$\frac{d}{d\phi}(\omega_2-\omega_1)=0,$$

where θ, ϕ are the usual angular coordinates: so that $\omega_2-\omega_1$ is constant over the surface, and may be taken as zero

...............(129).

Also the normal magnetic polarisation is continuous.

$$\therefore \{xL+yM+zN\}_1^2=0,$$

i.e. $\mathbf{rH}_1=\mathbf{rH}_2$, if $\mathbf{r}\equiv(x,\ y,\ z)$.

$$\therefore \mathbf{rH}_1''+\frac{K}{V}\mathbf{r}[(\mathbf{u}''-\alpha\mathbf{u})\mathbf{E}_1]=\mathbf{rH}_2''+\frac{1}{V}\mathbf{r}[\mathbf{u}''\mathbf{E}_2];$$

$$\therefore \mathbf{r}\nabla\omega_1-\frac{K}{V}(\mathbf{u}''-\alpha\mathbf{u})[\mathbf{rE}_1]=\mathbf{r}\nabla\omega_2-\frac{1}{V}\mathbf{u}''[\mathbf{rE}_2].$$

Now as the tangential component of $\mathbf{E}'$ is continuous and $\mathbf{E}'=\mathbf{E}$,

$\therefore$ $\mathbf{E}_1-\mathbf{E}_2$ is radial in direction;

$$\therefore \quad [\mathbf{rE}_1] = [\mathbf{rE}_2];$$

$$\therefore \quad \mathbf{r}\nabla(\omega_2 - \omega_1) + \frac{1}{V}\{(K-1)\,\mathbf{u}'' - \alpha K\mathbf{u}\}\,[\mathbf{rE}_1] = 0,$$

and
$$\mathbf{E}_1 = \left(0,\ 0,\ \frac{3}{K+2}\right),$$

so that
$$[\mathbf{rE}_1] = \frac{3}{K+2}(y,\ -x,\ 0):$$

$$\therefore \ \mathbf{r}\nabla(\omega_2 - \omega_1) + \frac{3}{(K+2)V}$$
$$\{(K-1)\,yA - \alpha Ky\,(A - ny) - (K-1)\,xB + \alpha Kx\,(B + nx)\} = 0.$$

Now at the surface

$$\mathbf{r}\nabla = x\frac{d}{dx} + y\frac{d}{dy} + z\frac{d}{dz} = a\frac{d}{dr},$$

when the coordinates are r, θ, ϕ.

$$\therefore \quad a\frac{d}{dr}(\omega_2 - \omega_1) + \frac{3}{(K+2)V}$$
$$\{(\beta K - 1)(Ay - Bx) + \alpha Kn\,(x^2 + y^2)\} = 0;$$

$\therefore$ observing that $\beta K - 1 = \alpha$,

$$\frac{d}{dr}(\omega_2 - \omega_1) + \frac{\alpha}{(K+2)Va}$$
$$\{Kn\,(2a^2 + x^2 + y^2 - 2z^2) + 3\,(Ay - Bx)\} = 0 \quad \text{...(130)}.$$

We have found also, in (129), that $\omega_2 = \omega_1$ at the surface, and that

$$\nabla^2\omega_1 = \frac{6K}{K+2}\frac{\alpha n}{V} \quad \text{..................(127)},$$

$$\nabla^2\omega_2 = 0 \quad \text{..............................(128)}.$$

If we had (127), (128), (129) and $\dfrac{d\omega_2}{dr} = \dfrac{d\omega_1}{dr}$ at the surface we should deduce

$$\left.\begin{aligned} \omega_1 &= -\frac{K}{K+2}\frac{\alpha n}{V}(3a^2 - r^2) \\ \omega_2 &= -\frac{2K}{K+2}\frac{\alpha n}{V}\frac{a^3}{r} \end{aligned}\right\} \quad \text{............(131)}.$$

But inasmuch as we have condition (130) instead of

$$\frac{d\omega_2}{dr} = \frac{d\omega_1}{dr},$$

we must superpose a distribution of ω satisfying $\nabla^2\omega_1 = 0$, and conditions (128), (129), (130); i.e. we must add to (131)

$$\omega_1 = \frac{\alpha}{(K+2)V}[Kn\{2a^2 + (x^2 + y^2 - 2z^2)/5\} + \{Ay - Bx\}],$$

$$\omega_2 = \frac{\alpha}{(K+2)V}[Kn\{2a^3/r + (x^2 + y^2 - 2z^2)a^5/5r^5\} + \{Ay - Bx\}a^3/r^3].$$

Hence in all we find that the external magnetic force $\mathbf{H}''$ has components which may be derived from a magnetic potential Ω_2, where

$$\Omega_2 = -\omega_2 = \frac{K-1}{(K+1)(K+2)V}\left\{Kn(2z^2 - x^2 - y^2)\frac{a^5}{5r^5} + (Bx - Ay)\frac{a^3}{r^3}\right\}.$$

The term in $(2z^2 - x^2 - y^2)$ may be regarded as due to a current sheet of the character already considered in § 26; the flow is in circles for which θ is constant, and varies as $\sin\theta\cos\theta$.

The term in $Bx - Ay$ would be produced by uniform internal magnetisation in the direction $(B, -A, 0)$ or by the equivalent sheet formed by constant currents in circles whose planes are equidistant. The axis of these circles would be the direction $(B, -A, 0)$, i.e. that of a diameter in the equatorial plane at right angles to (A, B, C), the velocity of the earth.

Close to the sphere the ratio which the forces due to the rotation bear to those due to the translation will be comparable with $an/(A^2 + B^2)^{\frac{1}{2}}$. If the velocity of the earth with respect to the "free ether" which we have taken as the origin of reference, be comparable with the velocity of the earth with reference to the sun, and if we take a sphere of 10 cm. in diameter revolving 100 times a second, this ratio is about 1/1000: thus the forces due to the translation are enormously the greater.

Röntgen's disc was 10 cm. in diam. and ·35 cm. thick, but he failed to observe any effects corresponding to the terms in

A, B from changing the sign of the electric field when the sphere was not rotating. We may infer either that the velocity of the "free ether" near the sphere is sensibly equal to that of the earth (i.e. that the earth carries the ether with it and that the explanation of stellar aberration has still to be found), or that the hypothesis of § 58 does not correspond with the facts.

That the latter is the more natural view to adopt, follows, I think, from the arbitrary nature of the assumptions involved in § 58 by comparison with those of the alternative theory of § 14, that the polarisation of a material medium is molecular and that it drifts through a stationary ether. Moreover still further hypothesis would be necessary if we attempted to explain the process by which according to § 58 small bodies must be regarded as carrying the ether with them with a velocity $(K-1)/(K+1)$ times their own, while there is to be no relative motion between the ether and the earth: the explanation of stress would also present new difficulties.

The evidence afforded by chemical and magneto-optical phenomena in favour of an interpretation in terms of ions is extremely strong; and a comparison of the results of Part IV with those of Parts II, III will give that evidence additional weight.

Printed in the United States
By Bookmasters